Progress in Sensory Physiology

Editors: H. Autrum, D. Ottoson,
E.R. Perl, R.F. Schmidt, H. Shimazu, W.D. Willis
Editor-in-Chief: D. Ottoson

Volume 4
1983. 41 figures. V, 118 pages
ISBN 3-540-12498-5

Contents: *N. Mei:* Sensory Structures in the Viscera. – *G.R. Martin:* Schematic Eye Models in Vertebrates. – *A. Gallego:* Organization of the Outer Plexiform Layer of the Tetrapoda Retina.

Volume 5
1985. 83 figures. V, 174 pages
ISBN 3-540-15339-X

Contents: *R.C. Hardie:* Functional Organization of the Fly Retina. – *H. Machemer, J.W. Deitmer:* Mechanoreception in Ciliates. – *M. Lindauer, H. Martin:* The Biological Significance of the Earth's Magnetic Field. – *H. Bleckmann:* Perception of Water Surface Waves: How Surface Waves are Used for Prey Identification Communications.

Volume 6
1986. 86 figures. V, 227 pages
ISBN 3-540-15340-3

Contents: *T. Sato:* Receptor Potential in Rat Taste Cells. – *K.B. Døving:* Functional Properties of the Fish Olfactory System. – *J.A. Coles:* Homeostasis of Extracellular Fluid in Retinas of Invertebrates and Vertebrates. – *S. Mense:* Slowly Conducting Afferent Fibers from Deep Tissues: Neurobiological Properties and Central Nervous Actions.

Volume 7
D.R. Irvine
The Auditory Brainstem
1986. 60 figures. XI, 143 pages
ISBN 3-540-16299-2

Contents: Introduction and Plan of the Review. – The Performance of the Auditory Central Processor. – Auditory Nerve Input to the Central Processor. – Cochlear Nucleus: Anatomy and Physiology. – Superior Olivary Complex: Anatomy and Physiology. – Nuclei of Lateral Lemniscus: Anatomy and Physiology. – The Auditory Midbrain: Anatomy and Physiology. – Auditory Brainstem Processing: Integration and Conclusions.

Progress in

Sensory Physiology 8

Editors: H. Autrum D. Ottoson
E. R. Perl R. F. Schmidt H. Shimazu W. D. Willis

Editor-in-Chief: D. Ottoson

With Contributions by
W. Skrandies J. Le Magnen A. Faurion

Springer-Verlag
Berlin Heidelberg New York
London Paris Tokyo

With 56 Figures and 16 Tables

ISBN-13: 978-3-642-71062-9 e-ISBN-13: 978-3-642-71060-5
DOI: 10.1007/978-3-642-71060-5

The Library of Congress has cataloged this serial publication as follows:
Progress in sensory physiology. – 1- – Berlin; New York: Springer-Verlag, 1981 – v.: ill.; 25 cm. Irregular. Editor: D. Ottoson. ISSN 0721-9156 = Progress in sensory physiology. 1. Senses and sensation-Collected works. 2. Sense-organs-Collected works. 3. Neurophysiology-Collected works. I. Ottoson, David, 1918 –.
[DNLM: 1. Neurophysiology-periodicals. 2. Sensation-physiology-periodicals.
W1 PR681G] QP431.P73 599'.01'8-dc19 85-647743 AACR 2 MARC-S [8509]

Softcover reprint of the hardcover 1st edition 1987

Typesetting: K+V Fotosatz GmbH, Beerfelden

2121/3140-543210

Contents

The Upper and Lower Visual Field of Man: Electrophysiological and Functional Differences*

W. Skrandies

Max-Planck-Institut für Physiologische und Klinische Forschung,
W. G. Kerckhoff-Institut, 6350 Bad Nauheim, FRG

* To Annemarie and Gernot

Progress in Sensory Physiology. Vol 8

We are all aware of the fact that science has become more and more impatient and competitive. New experiments and new techniques may open new ways of understanding the nervous system. However, their success is based entirely on the intelligent and responsible use of them, which implies a thorough knowledge of what has been learned from the serious work of the past. The questions remain the same, and the techniques themselves should not be overvalued. Rather, the knowledge and results presented ... should be judged by the intelligence of the question, the seriousness of the approach, and the rigor of the conclusion, no matter whether the technique was traditional or modern.

O. D. Creutzfeldt
(Words of Welcome, Seventh European Neuroscience Congress, Hamburg, 1983)

1 Introduction: The Divided Outer World

Basic sensory functions depend on physical stimulus characteristics, physiological perceptual mechanisms, the state of the nervous system at the time of stimulus arrival, and also on the exact location of the stimulus on the receptor plane. In the somatosensory system, for example, the spatial resolution for mechanical stimuli differs according to whether they are applied to the fingertips or to the arm area, and the visual system shows pronounced functional differences between the center and the periphery of the retina, related to the different local packing densities of the photoreceptors and their different connections to the retinal ganglion cells.

The density of peripheral receptor and nerve cells has a direct relation to the cortical representation in a given sensory modality. In the somatosensory system a distorted representation of the body surface results in the homunculus, where relatively large cortical areas are devoted to small portions of the body like the tongue or the fingers, while other, larger areas of the body surface have only small cortical representations. In a similar way, the retinotopic projection of the visual field onto the visual cortex appears to be weighted by its functional importance and the anatomical wiring in the retina. Due to the cortical magnification factor (Whitteridge 1973), the central areas of the retina have a much larger cortical representation than the corresponding peripheral areas (Holmes 1945). Information processing in the sensory systems of the brain is performed in a complex manner, partly sequential, partly parallel. Differences in processing of incoming information depend not only on physical stimulus attributes but also on the external location of the stimulus to the right or to the left of the subjective vertical. Because of the basic wiring of most mammalian nervous systems, including man's, primary sensory information of all modalities is directed to either the left or right hemisphere of the cerebral cortex. Although there exists some overlap, in general, stimuli which appear on the left of the subjective vertical are projected to the right cerebral hemisphere while stimuli on the right are projected to the left hemisphere. Similarly, the motor system is dominated by contralateral connections, and for more specialized purposes like writing or tool handling one hand is sufficient and one side of the brain may come to be preferred. The question of how the brain achieves its integrative task of uniting the halves of the perceptual world and enabling coordinated motor activity of twin organs has prompted much research in the neurosciences.

Since Broca's (1861) and Wernicke's (1874) work on aphasia, functional differences in language abilities between the two hemispheres have been known. This led to the concept of lateralized speech centers in the brain. Starting with Sperry's experiments on "split-brain" patients (Sperry 1964, 1968), other differences in general information-processing strategies which cannot simply be attributed to different language capabilities of the left and right halves of the brain were established, and a vast scientific literature has since accumulated (e. g., Gazzaniga 1970; Dimond and Beaumont 1974). The basic concept that the left hemisphere is superior in processing sequential, abstract input, while the right hemisphere is better in dealing with holistic, gestalt-type information processing, including the recognition of faces or spatial orientation (cf Jeeves 1984; Grüsser 1984), has now attained general acceptance. These results are based on experimental work in various sensory modalities.

In the visual system, functional differences between different regions of the visual field cannot be discussed simply in terms of a left/right distinction, because the left visual field projects onto the nasal retina of the left eye but onto the temporal parts of the right eye, whereas stimuli in the right visual field are processed by the nasal retina of the right eye and the temporal retina of the left eye. Anatomical and physiological left/right differences in the visual system can be located only beyond the optic chiasm, where the nerve fibers originating in the nasal retina cross to the contralateral cerebral hemisphere. The functional regional differences which occur in visual acuity (Wertheim 1894; see Sects. 5.2 and 6.2.2.3) are attributed to the different densities of photoreceptor distribution in the retina and appear also to be related to the relative importance of the occurrence of visual stimuli in the center or the periphery of the visual field. In addition, the regional distribution of temporal sensitivity of the visual system shows a systematic difference between the nasal and temporal parts of the retina which does not simply reflect general differences between the right and left visual fields or between the functional capability of the left and right cerebral hemispheres (see Sects. 5.1 and 6.2.2.2).

Eye movement recordings during fixation suggest that the vertical meridian also plays some functional role in vision. Constant movement of the image on the retina is an important prerequisite for normal visual function. Human psychophysical data suggest that eye movements are related to the fact that high spatial sensitivity of the retina is confined to a very small area of about 20′ arc diameter in the center of the fovea (Alpern 1972). The perception of a visual scene is reconstructed from a sequence of sensory data obtained by scanning over different positions in the visual field. In addition, continuous eye movements are important for perception, since stabilized retinal images result in the disappearance of the visual percepts (Yarbus 1967; Ditchburn 1973). During visual fixation of an object, drifts and saccades occur with a directional specificity: Nachmias (1959) found that 63% of all saccades go off at between 50° and 90° (i. e., toward the upper octant of the visual field). A bivariate normal distribution of eye movement directions was computed which was tilted toward the vertical meridian (see Fig. 1A). The results of Bennet-Clark (1964) and of Yarbus (1967) on eye movements during continuous fixation of a visual target are in agreement, as illustrated in Fig. 1B and C: The eye fixation fields are not restricted to horizontal movements but contain a pronounced vertical component.

This suggests that the vertical direction is also important for visual perception. The results which I will discuss show that functional regional differences in the visual system are not confined to the horizontal meridian reflecting cerebral hemispheric specialization in terms of a left/right distinction; there exist, in addition, related differences between retinal areas revealed by electrophysiological and psychophysical experiments. These data suggest that different characteristics of the perceptual mechanisms are activated by stimuli in the upper and lower halves of the visual world. I will thus not consider the well-known, widely propagated, and presently popularly accepted differences in higher information processing presumably located in the two halves of the cerebral cortex, but rather review anatomical, physiological, and functional differences between the upper and the lower hemiretinal systems.

Some of the earliest quantitative experimental work on visual perception dealt with visual acuity. Aubert and Förster (1857) and Landolt and Hummelsheim (1904) investigated the regional differences in spatial visual resolution, and their results showed that visual acuity is higher for upper than for lower retinal stimuli. This suggests a fundamental functional system difference (see also Sect. 6). In the course of electrophysiological studies in man (Eason et al. 1967; Lehmann et al. 1977; Lehmann and Skrandies 1979) it has been observed that stimulation of upper retinal areas evokes electrical brain activity with shorter component latencies than stimulation of lower retinal areas. These findings lead to the question of whether the functional (visual acuity) difference is related to the electrophysiological difference reflected in component latency. In addition, it appears of interest to investigate whether only some isolated visual functions differ in the upper and lower visual fields, or whether there is a global distinction in visual processing capability.

Various aspects of perceptual efficiency can be examined psychophysically, as illustrated in Sect. 5, while electrophysiological methods offer the possibility of analyzing retinal and cortical activity independently. Such electrophysiological data will help to establish at what stage of the visual system the difference in per-

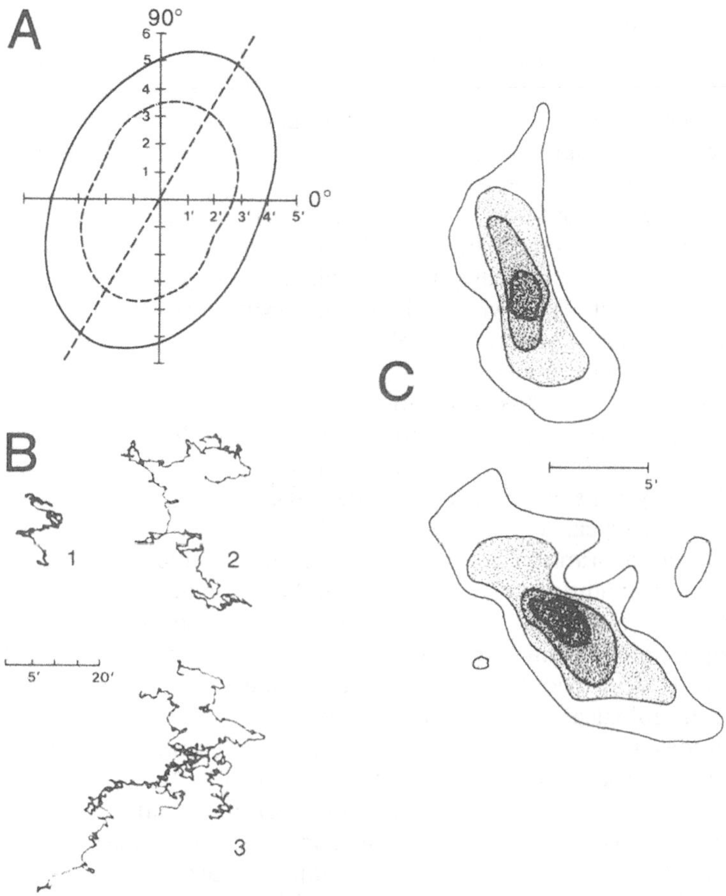

Fig. 1A–C. Distribution of eye movements during fixation of a steady point. **A** Direction of the visual axis, showing an approximately bivariate distribution. Any point of the retinal image will lie in the region of the solid curve 68% of the time during a fixation interval of 30 s. (Data from Nachmias 1959, Fig. 6; redrawn after Ditchburn 1973) **B** Eye movements during fixation for *1.* 10 s, *2.* 30 s, or *3.* 60 s. (Redrawn after Yarbus 1967, Fig. 54) **C** Contour maps of the direction of the visual axis for two subjects showing the area scanned during a fixation period of 10 s. 100% of all eye movements lie within the outermost border of the fixation fields; 25% lie within the innermost contour (Redrawn after Bennet-Clark 1964, Figs. 5 and 6)

formance is initiated. Differences in the processing of the upper and lower retinal visual stimuli could be explained by a general superiority of the upper over the lower hemiretinal system or alternatively by local physiological differences. Evoked potential data are obtained only at one of the last steps in the chain of visual information processing, and thus must be complemented by the analysis of activity at earlier stages in the visual system in order to localize the origin of the functional differences.

The correct interpretation of evoked potential data depends on the use of unambiguous recording and data analysis techniques in order that we may arrive at physiologically meaningful conclusions from scalp-recorded electrical brain activity. Such methods have been developed as topographical techniques during recent years (Lehmann 1977; Lehmann and Skrandies 1980, 1984; Skrandies and Lehmann 1982a, b), and these techniques have been used successfully to study electrophysiological differences between the upper and lower hemiretinal systems in man, as illustrated in Sect. 3.

I will first discuss how scalp-recorded brain activity can be used to gain insight into complex electrical brain mechanisms in spite of some inherent shortcomings of the "indirect" recording method which is the only possible one in intact human subjects (Sect. 2). It will be shown that evoked potential recordings in man may yield information which complements the results of "direct" intracranial recordings in animals. Evoked potential data reflect the mass activity of large neuronal populations, and thus allow conclusions on global functional differences of larger brain areas which may be missed in selective single unit recordings (see Sect. 2.2). Various experiments which have been performed comparing the upper and lower retinal systems in man will be described below. First, I will report on electrophysiological differences in evoked potentials between upper and lower hemiretinal stimulation obtained in different subject populations under various stimulus conditions (Sect. 3). The implications of the results for the clinical application of evoked potentials will be demonstrated in Sect. 3.3. Then electrophysiological investigations of the different stages of the human visual system will be described (Sect. 4), and an attempt will be made to relate the electrophysiological findings (which are usually discussed only in formal terms of amplitudes, latencies, or scalp topography) to functional differences between the upper and lower hemiretinal systems. Prompted by the electrophysiological results, a series of studies has been made investigating visual functions such as temporal sensitivity, contrast sensitivity, and discrimination of meaningful stimuli (Sect. 5). These functional aspects appear of major importance, since even the most significant finding is not of much interest if its interpretation remains restricted to a small and highly specialized field of research and is not paralleled by behavioral concomitants.

Evidence on differences between the upper and lower retinal systems is scattered through the literature, and in most cases differences between upper and lower retinal areas have not been studied systematically. A review of the pertinent data from neuroanatomy, physiology, psychophysics, psychology, and ophthalmology will be given in Sect. 6.

2 Analysis and Origin of Visual Evoked Potential Components[1]

2.1 Topographical Analysis of Scalp-Recorded Brain Activity

In the early 1870s the English physiologist R. Caton started to investigate the electrical activity of the brain, and in 1875 he presented his results obtained in rabbits and monkeys to the British Medical Association. From these experiments it

[1] A related short account of some of these ideas illustrating different sets of data has been given (Skrandies 1986a) elsewhere

became clear that electrical brain activity can be divided into spontaneous activity and evoked activity: "In every brain hitherto examined, the galvanometer has indicated the existence of electric currents. ... Impressions through the senses were found to influence the currents of certain areas ..." (Caton 1875). The same is true for the electrical activity of neural populations in the human central nervous system which can be recorded from the intact scalp of man: the electroencephalogram (EEG) reflects spontaneous activity, as first discovered by Berger (1929), while evoked potentials are in general elicited by adequate stimulation of sensory organs.

Evoked brain potentials reflect the synchronous activity of neuronal populations whose geometry in the brain produces an electrical field which can be recorded from scalp electrodes. Because of volume conduction this electrical field can be detected over the entire scalp, and conventional recordings of potential differences between two electrodes reflect only the potential gradient along the line between the recording sites. As will be shown in this section, the analysis of such potential waveshapes should be replaced by an adequate quantitative treatment of the electrical data as scalp field distributions. Recording of evoked potentials is a noninvasive method with high time resolution (down to milliseconds) and high sensitivity to spontaneous and induced changes of the state of the brain. Although the mechanisms of the electrogenesis of scalp-recorded brain activity are not clear in detail, studying its covariation under experimental conditions appears fruitful. A high correlation between the firing probability of single cortical cells and the evoked potential recorded near the cell body was reported by Fox and O'Brien (1965). Creutzfeldt and Kuhnt (1967) concluded from their studies on cats that evoked potentials are composed in inhibitory and excitatory postsynaptic potentials as well as of synchronous fiber activity. Thus, a component of the evoked potential may reasonably be defined as the occurrence of synchronous activation of a large neural population. Due to the fact that the human cortex is strongly folded, electrical fields generated by oppositely oriented neuronal generators will cancel. In spite of this problem in all sensory modalities, there are scalp-recorded components that are very consistent across subjects. An evoked component may be recorded from the intact human scalp:

1. If it has a fixed temporal relation to the occurrence of a physical stimulus or to some internal stimulus-processing mechanisms [which elicits a so-called endogenous component; see Donchin et al. (1978) and Skrandies et al. (1984)]
2. If there is a population of neurons in the brain which is activated synchronously
3. If the neuronal structures involved produce an open electrical field as a net response

2.1.1 Why Topography?

Topographical analysis of EEG and evoked potential data is in general considered a tool to define the intracranial sources of scalp-recorded electrical brain activity. The methods of localizing the sources of electrical activity are based on the work of von Helmholtz (1853). It is a physical fact that the scalp potential distribution can be calculated when the instantaneous configuration, location, and orientation of the activated intracranial neural populations are known [the so-called forward problem; cf Wilson and Baley (1950)]. But it is important to keep in mind that the reverse is not true: scalp field configurations do not allow unequivocal predictions of the underlying intracranial sources. Attempts to solve this "inverse" problem, i. e., how to calculate the sources of potentials in a conductive medium when the potential field on the surface is given, have no clear solution. Therefore, computations to localize model dipoles always have to rest on certain assumptions regarding conductivity, homogeneity, and geometry of intracranial media, and assumptions concerning the number, location, and extension of electrical model dipoles must also be made in order to obtain physiologically meaningful interpretations (Schneider 1972, 1974; Kavanagh et al. 1978; Sidman et al. 1978). Because of this "inverse" problem, and because of the fact that millions of individual neural generators are active at any given moment, the goal of intracranial generator localization presently appears of less practical value than claimed.

This chapter will discuss a different approach to the treatment of topographical data. I will show that scalp distributions are being investigated primarily not in order to identify intracranial sources of evoked brain activity, but because the scalp topography of the electrical fields of the brain characterizes brain electrical events unambiguously in terms of latency and scalp location. In addition, topographical data provide a basis for testing hypotheses concerning the processing mechanisms of different aspects of sensory information by the central nervous system. Irrespective of whether the exact intracranial sources of the scalp-recorded brain activity are known, the interpretation of scalp distribution data, combined with knowledge of the anatomy and physiology of sensory brain systems, can lead to physiologically meaningful conclusions. Comparison of scalp distributions evoked under different stimulus conditions gives direct information on the identity (or nonidentity) of the neural populations activated under these conditions. While identical distributions may or may not be generated by different neural populations, nonidentical scalp fields must be caused by different neural generator mechanisms. In the present section I will discuss how scalp-recorded brain activity can be used to investigate complex electrical brain mechanisms independent of the basic shortcomings of the indirect (i. e., extracranial) recording method which is in general the only one possible in human subjects.

Section 2.1.2 will show how quantitative measures of electrical brain activity are extracted without the choice of a reference electrode. Subsequent sections will illustrate applications of these techniques to physiological questions concerning the differences between upper and lower retinal areas, which are revealed by evoked potential recordings under various stimulus conditions.

2.1.2 Some Basic Considerations

Evoked potential data is commonly analyzed in terms of components which are believed to be subsets of electrical brain activity related to different neural processing mechanisms. The prime aim of evoked potential studies is the establishment of relations between such components and experimental or physiological parameters. The problem of adequate definition of components has been extensively discussed in the literature. Techniques ranging from simple peak measurement to elaborate multivariate statistics have been proposed, and the relationship to underlying neuroanatomical structures has also been stressed:

> The definition of the components is an empirical event and their localization to sets of neurones requires a deep knowledge of the structure and dynamics of the region in which the event occurs, to the extent that the locations, patterns of spread, and time courses can be predicted for particular components. ... It is, to be sure, a long way from the olfactory bulb to the association neocortex, but I will maintain that unless you develop and use a theoretical basis for predicting waveforms from local neural anatomy, topology, and dynamics, you will be wallowing in principal components indefinitely (Freeman 1978, p. 507).

One should add that evoked potential analysis must not only consider the underlying neural structures, but has first of all to acknowledge that we are dealing not with sets of potential waveshapes but with electrical fields generated by the brain. Much controversy regarding evoked potential data originates from the ambiguous interpretation of waveshapes. Looking at potential fields can prevent one from "wallowing in components," and in the following sections I will illustrate quantitative methods which analyze evoked potential data adequately as electrical fields.

Scalp-recorded electrical brain activity reflects the mass response of large neural populations which spreads throughout the brain via volume conduction. This is illustrated by the auditory brain stem response: auditory stimuli are followed by potential components of very short latency (less than 10 ms) which can be recorded from the intact scalp of man, and there is direct evidence that these components reflect the electrical activity at the early processing stages of the auditory pathway at the level of the brain stem (Jewett et al. 1970; Starr and Achor 1975). Field potentials generated by the activation of optic nerve and optic tract fibers in the cat visual system can be recorded in widespread regions of the brain (Skrandies et al. 1978). Detailed analysis of field potentials in the retina is possible only in animal experiments by means of direct intraocular recording of mass activity and single unit responses (Fig. 2A). The mass activity of the Y-, X-, and W-ganglion cells of the cat retina following electrical stimulation of the optic tract may be discriminated at the optic disc (Fig. 2B), and the field potentials "stray" from the retina into the vitreous body of the eye. The field potential components persist even when the electrode is several millimeters in front of the retinal surface (Fig. 2C–E). In addition, a far-field potential generated by the Y-fibers entering the eye can still be seen over the temporal retina at 6 mm eccentricity (Fig. 2C). Similarly, the early component of the Y-fibers can be recorded in many brain regions that have no direct anatomical connection to the optic nerve or tract (Skrandies et al. 1978), and these data show that potential components recorded at a given point in the nervous system need not be generated at that site. Thus, any statement about the location of neural generator populations of components of evoked brain activity requires a solid experimental and physiological basis.

In the field of EEG the problem of the position of the reference electrode is considered crucial, and this topic has received much discussion in the literature. I will show in this section how the problem can be avoided and how surface-recorded EEG data can be interpreted unambiguously independent of the location of the reference electrode.

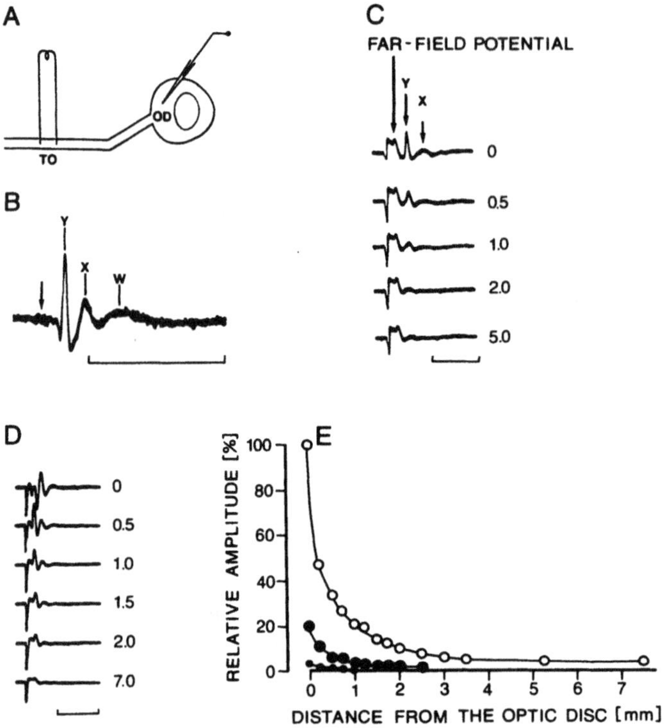

Fig. 2. A Experimental setup for the study of field potentials in the cat eye. Stimulating electrodes are stereotactically placed in both optic tracts (Horsley-Clarke coordinates: anterior +10, lateral 7). Single unit responses and field potentials were recorded from retinal ganglion cells. *OD*, optic disc; *TO* optic tract. Further details are described by Skrandies et al (1978) **B** Antidromic field potentials recorded at the optic disc after antidromic electrical stimulation of the optic tract. Three components can be identified as the mass responses of Y-, X-, and W-ganglion cells. *Arrow* indicates stimulus onset. **C** Field potentials over the temporal retina at 6 mm eccentricity after electrical stimulation of the optic tract recorded at various distances from the retina. *Numbers* indicate distance to the retinal surface in millimeters, and the amplitudes of the locally generated potential components decrease gradually while the amplitude of the far-field potential originating from the optic disc remains constant. The first sharp peak in the waveforms is the electrical stimulus artifact corresponding to the instant of stimulation. **D** Field potential components recorded over the optic disc when the recording electrode is pulled back from the retina. *Numbers* indicate distance to the retinal surface in millimeters. The first sharp peak in the waveforms is the electrical stimulus artifact corresponding to the instant of stimulation. **E** Amplitudes of field potential components over the optic disc as a function of distance to the retina. The three ganglion cell classes show a different amount of amplitude decrease. *Dots*, W-field potential; *closed circles*, X-field potential; *open circles*, Y-field potential. Time scales *(bars)* for all traces indicate 5 ms. (Data from Skrandies et al. 1978)

Electrical brain activity is recorded as potential differences between recording points on the scalp, and accordingly the position of both the reference and the recording electrode influences the waveform pattern of electrical activity. The potential waveforms are always ambiguous in terms of amplitude and phase of electrical events as analyzed over time, but the characteristics of the underlying

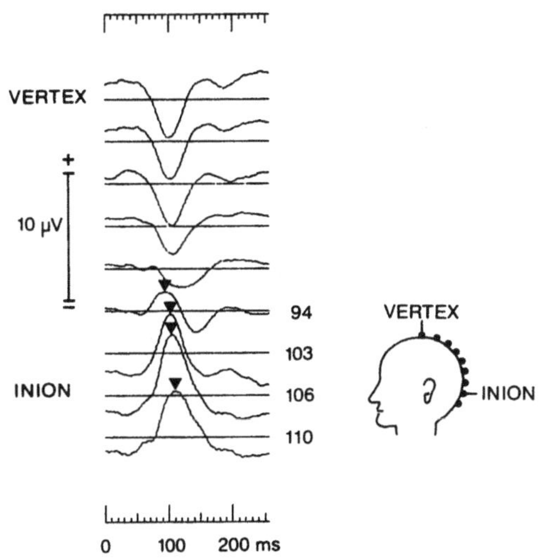

Fig. 3. Potential waveshapes evoked by a 30′ arc reversing checkerboard pattern presented with 62% contrast as a test field of 8.7°×11.4° arc to the upper hemiretina. *Arrowheads* and *numbers* indicate peak latencies (in ms) of the corresponding waveshape. The electrode array starts at the vertex and extends in regular steps to 2 cm below the inion (see head schema on the *right*). Data are referred to the average reference

electrical fields of the brain (and thus of the scalp fields) are constant and reference-independent at each moment in time. Accordingly, only electrical fields may be interpreted in a meaningful way (Lehmann 1977; Nunez 1981). Strict topographical analysis strategies must be applied to electrical scalp fields in order to avoid the problem of the reference electrode and ambiguous data interpretation (Lehmann and Skrandies 1980; Skrandies and Lehmann 1982a).

The inspection of evoked potential waveshapes may yield confusing results. Visual evoked potentials (VEPs) recorded from nine midline electrodes evoked by a checkerboard reversal stimulus presented to the upper hemiretina are shown in Fig. 3. These waveshapes are referred to the "average reference", i. e., to the mean potential of all electrodes in the field. Note that the major occipitally positive component shows different peak latencies in different recording channels, as indicated by the arrowheads in Fig. 3 (e. g., channel 6 has a peak latency of 94 ms; channel 7, 103 ms; channel 8, 106 ms; channel 9, 110 ms). These systematic changes in latency might be interpreted as evoked activity moving from anterior to posterior brain areas. A change of the location of the reference electrode makes the picture more complex. The identical data set recomputed against other reference sites is illustrated in Fig. 4A: On the left, the data of Fig. 3 are referred to the vertex reference, in the center an electrode 2.5 cm posterior to the vertex is used as reference, and on the right an electrode 5 cm posterior to the vertex is used as reference point. Latency differences occur in different recording channels, and there appear changes in latency depending on the reference chosen: for example, channel 8 has peak latencies of 103, 105 or 106 ms. The complete possible data set consists of the potential waveshapes referred to every other electrode of the array (i. e., for nine electrodes 72 different waveshapes are possible). All these waveshapes are illustrated in Fig. 4A–C. Evidently, this rather simple set of data presents an astonishing amount of information and confusion. It is important to keep in mind that we are dealing with identical data (the same data set has been recomputed) and that all statements about peak latencies are correct. There is no good reason to prefer one set of waveshapes shown in Fig. 4 over the others. This

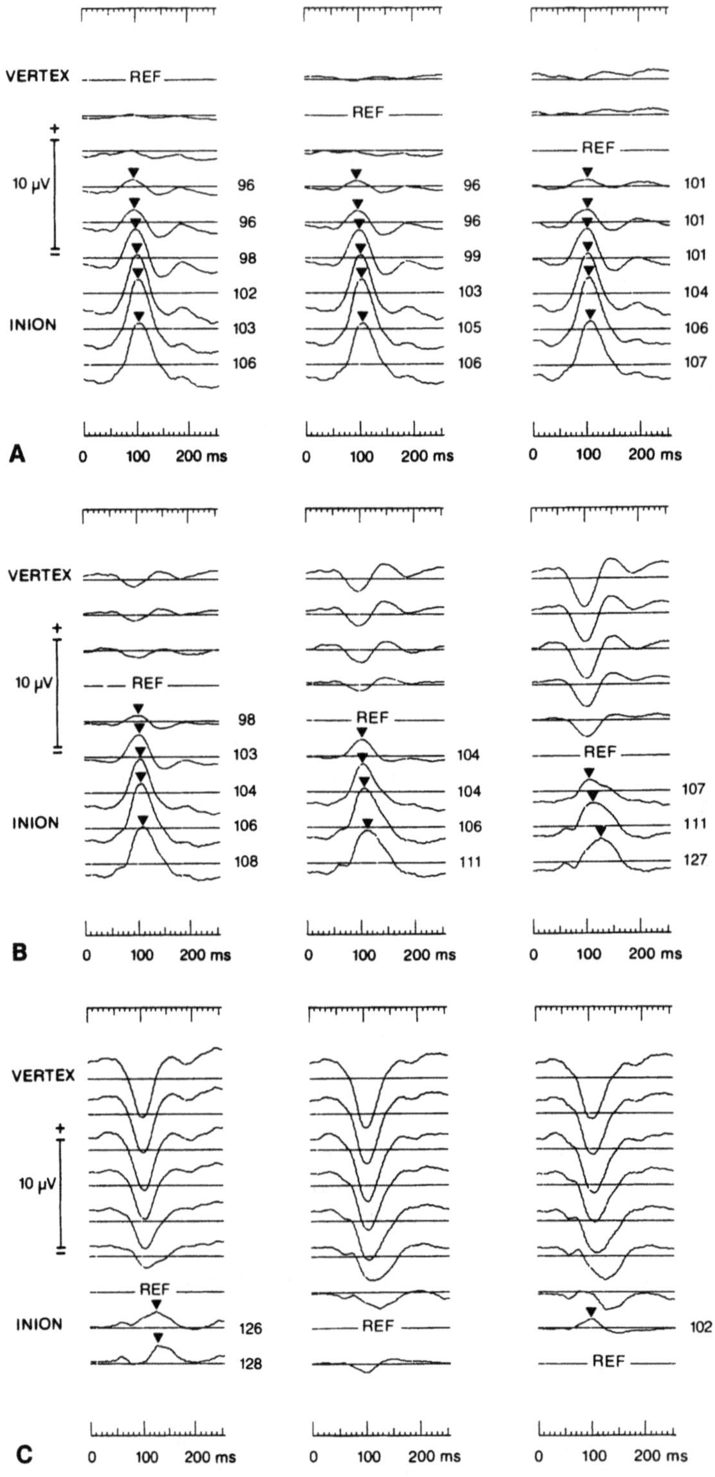

Fig. 4A–C. Data of Fig. 3 computed vs all possible reference sites. Each column shows the waveshape patterns when a given electrode is used as the reference. The reference channel (*REF*) shows a flat line. First electrode at the vertex (Cz), last electrode 2 cm below the inion. Consult head schema in Fig. 3 for electrode locations. *Arrowheads* and numbers indicate peak latencies (in ms) of the corresponding waveshape

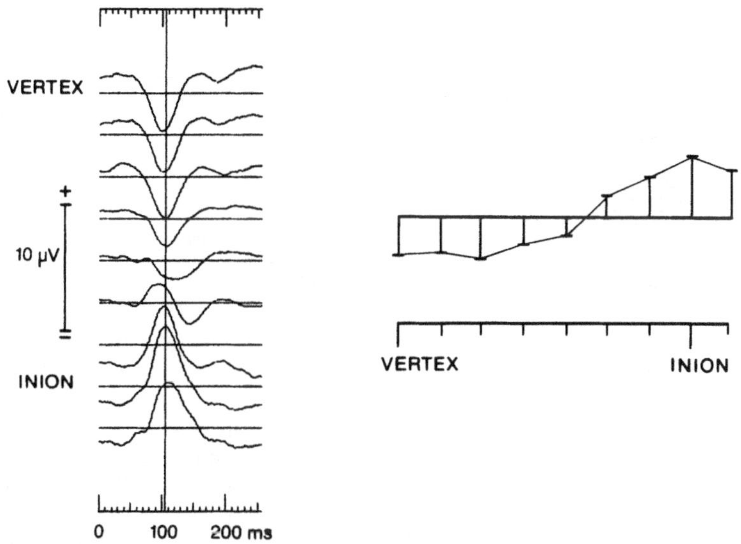

Fig. 5. Construction of a topographical potential profile at 106 ms latency. Amplitudes measured in each channel are plotted as a function of electrode position. The *zero line* corresponds to the value of the average reference. Changing the reference site shifts the zero baseline up or down, but leaves the shape (i.e., the location of maxima and minima, and the gradients) of the profile unchanged

dilemma of waveshape variability and ambiguous data intepretation can be overcome only by an adequate analysis technique which is independent of the reference electrode.

In contrast to what is suggested by Figs. 3 and 4, electrical brain activity does not move over larger distances on the scalp and different component latencies in different recording channels are not observed. This becomes evident when the data are treated as an electrical field, and Fig. 5 displays the data at 106 ms latency as a potential profile. This is simply a different arrangement of the data: the amplitude values at 106 ms are measured in each channel, and are entered at their respective electrode positions (Fig. 5, right) resulting in a potential profile at the respective instant. When using two-dimensional electrode arrays, potential maps are the adequate topographical display of the electrical brain activity (e. g., Figs. 10, 23, 25). The major feature of the profile in Fig. 5 is an occipital positive maximum surrounded by steep gradients, while over anterior areas a flat distribution prevails. Most importantly, a change of the reference point will not change the shape of the potential profile, since a different reference would simply change the position of the zero potential level and thus the labeling of the potential values, while the potential maxima and minima and gradients within the field remained unaffected.

The information of Fig. 3 is topographically displayed in Fig. 6 as a series of potential profiles between 4 and 256 ms latency. This display contains the same amount of information as the waveshapes but stresses the topographical aspect of the data.

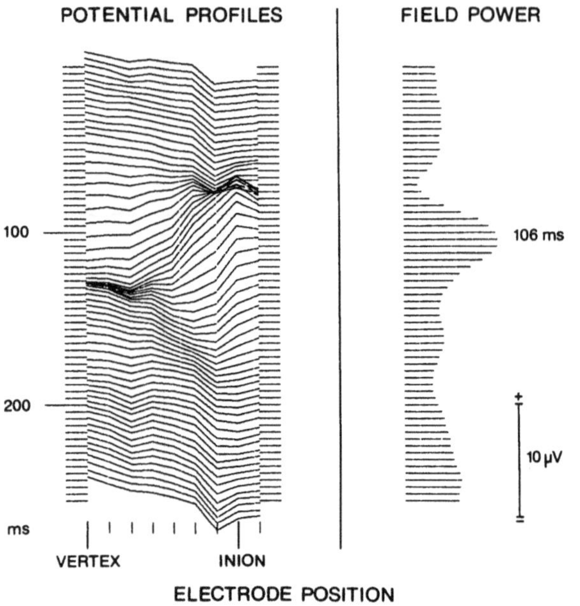

Fig. 6. Series of 64 consecutive potential profiles between 4 and 256 ms (*left*), and global field power (*right*) computed every 4 ms (same data as Figs. 3 and 4). The time axis runs from top to bottom, the anterior-posterior electrode positions from left to right. Positive potential values upward

It appears reasonable to define the latency of an evoked component as the occurrence time of maximal activity in the scalp field reflecting the time of synchronous activation of a maximal number of neurons. In order to quantify the electrical activity of each scalp field, we have proposed a measure of global "field power" that is computed as the root mean square of all possible potential differences in the field, or a related measure of "potential range" (i. e., the difference between the maximal and minimal potential values obtained, irrespective of their scalp location) which yields similar results (Lehmann and Skrandies 1980; Skrandies and Lehmann 1982a). These measures quantitatively determine the amount of electrical activity at every instant independent of the reference electrode, and pronounced scalp fields with high peaks and troughs and steep gradients will result in high field power. Times of maximal field power define the occurrence of a component. In Fig. 6 global field power is plotted on the right side as a function of time adjacent to the corresponding potential profiles. The field power is low at early latency times but increases around 100 ms, reaching a maximum at 106 ms. This instant defines the latency of the evoked component. Afterward, field power decreases again (see Fig. 6).

Topographical changes over time can be analyzed in detail by plotting the potential profiles at successive points in time, as illustrated in Fig. 7. Contrary to what was suggested by the potential waveshapes in Fig. 3, it is obvious that there is no change in the scalp location of the occipital component, because between 90 and 120 ms the peak remains at the same electrode and only the potential gradients change.

After determining component latencies, the next step is the analysis of the scalp field features at these latency points in terms of scalp locations of the maxima, minima, and potential gradients. These derived measures are by definition reference-independent (i. e., the location of a maximum or a minimum does not

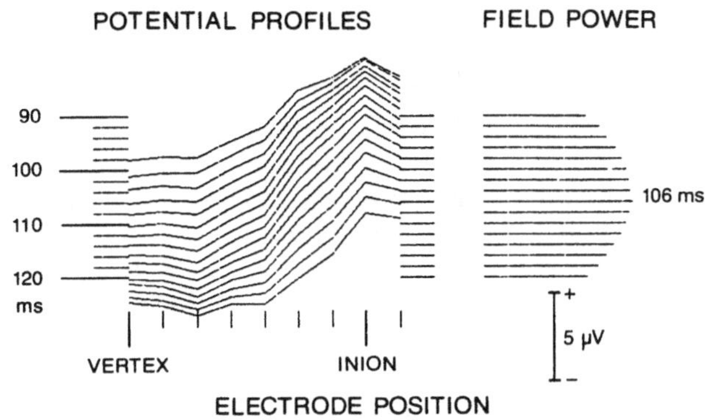

Fig. 7. Enlarged display of part of the profile series of Fig. 6, between 90 and 120 ms. Note that there is no change in component location as was suggested by the potential waveshape pattern of the same data in Fig. 3

change when the reference is changed). Different subject populations or different experimental conditions may then be compared statistically using these extracted values.

This topographical analysis strategy defines evoked components independent of the reference point, and treats electrical scalp field data unambiguously. Thus, scalp field data can be related to intracranial mechanisms in a physiologically meaningful way. This methodological approach is used in the present text to analyze scalp-recorded electrical data.

2.2 Origin of Visual Evoked Potential Components

In most visual evoked potential (VEP) studies the P100 component – an occipitally positive component of approximately 100 ms latency – is evaluated, since it is very reliable and shows low intersubject variability. On the basis of its long latency and its scalp distribution elicited by lateralized half-field stimulation (Skrandies and Lehmann 1982b), this component is presumably of cortical origin. Further support for a cortical origin of the P100 component was obtained by analyzing components evoked by patterns with contrast borders and comparable stereoscopic checkerboard patterns which depend on binocular fusion in the visual cortex.

A prerequisite for the analysis of the origin of VEP components is knowledge of the underlying neural processing steps in the visual system. Dynamic random dot stereograms (RDS) which do not contain monocular cues evoke potentials of cortical origin (Lehmann and Julesz 1978), since in the human visual system the input of the two eyes remains separated until the level of the visual cortex. The binocular interactions crucial for stereoscopic depth perception must be cortical functions. Thus, the "imaginary cyclopean eye" (von Helmholtz 1910; English edition Southall 1962) or the "cyclopean retina" (Julesz 1971) are operationalized in the visual cortex. Purely cortical neuronal generating mechanisms were studied by recording the brain's electrical activity evoked by dynamic RDS stimuli, skipping operationally all processing stages prior to the activation of cortical binocular disparity neurons. The comparison of a stereoscopic stimulus condition with a similar two-dimensional stimulus condition gives information on the similarity of the neural processing of these two classes of stimuli (Skrandies and Vomberg 1985).

In 15 subjects evoked potentials were recorded from a midline row of nine electrodes equidistantly spaced between the vertex and 5 mm above the inion (see inset in Fig. 8). Stereoscopic checkerboard reversal patterns composed of random dots were shown to each eye separately every 20 ms using a special visual pattern generator (for details see Skrandies and Vomberg 1985). This stereoscopic checkerboard was reversed every 256 ms, and the change served as the stereoscopic stimulus while monocular processing stages received only dynamic noise.
In order to assure that the RDS stimulus did not contain monocular cues, evoked brain activity was also measured with monocular stimulation. In this conditon evoked potentials did not exceed the noise level. This is in line with earlier findings that in stereoblind patients no stereoscopically evoked brain activity can be recorded (Vomberg and Skrandies 1985). Two-dimensional black and white checkerboard reversal stimuli presented binocularly with the same mean luminance and high contrast were used for comparison. Potential profiles were analyzed, component latency was determined by the occurrence of the maximal field power computed in a time window between 80 and 160 ms, and component locations were defined as the electrode position of maximal voltage at this latency (see Sect. 2.1).

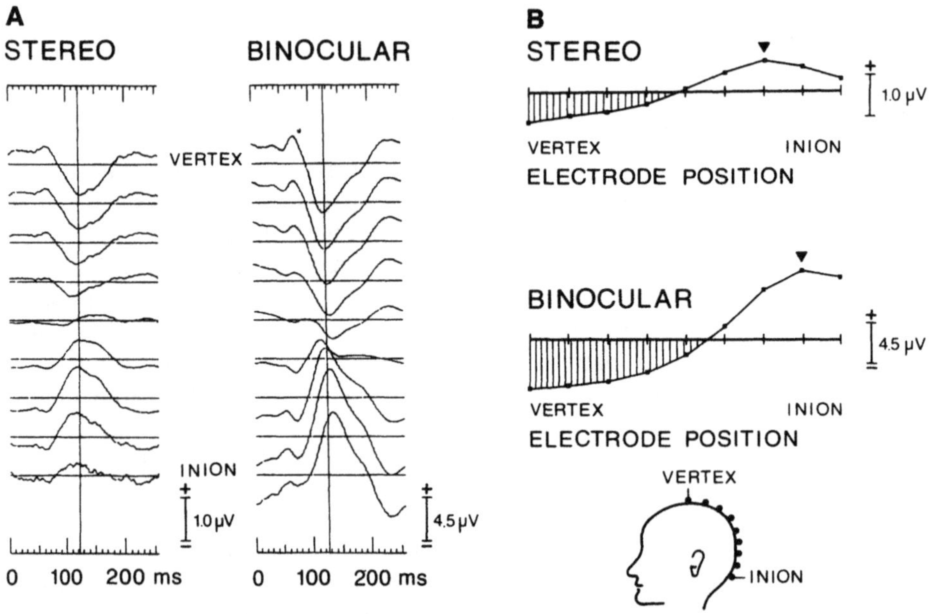

Fig. 8. A Grand mean potential waveshapes of 15 subjects evoked by stereoscopic stimuli or comparable binocular contrast stimuli. Data were recorded from a midline row of nine electrodes equidistantly spaced between the vertex and 5 mm above the inion, and are referred to the average reference. The occipitally positive component which was determined by the occurrence of maximal field power is marked by a *vertical line*. **B** Grand mean potential profiles at component latencies. The data of **A** are displayed as scalp field distributions. The location of the maximal positivity is marked by *arrowheads*. Note larger amplitudes of components evoked by contrast stimuli and the more anteriorly located potential maximum of the stereoscopically evoked component. Head schema shows electrode positions

The configuration of the potential waveshapes was similar for stereoscopic and binocular contrast stimuli: In both stimulus conditions an occipital positivity occurs about 120 ms after the stimulus (Fig. 8A). In the subject population component latencies were not different, whereas amplitudes of the RDS-evoked components were significantly smaller than those of the binocular contrast VEPs. Figure 8B shows the potential profiles at their respective component latencies. For both stimulus conditions a positive, occipital maximum is seen. This maximum was located more anteriorly for stereoscopic stimuli (mean 1.8 cm) in the subject population, and the difference in component location was statistically significant ($t = 3.59$, $p < 0.0025$).

Smaller amplitudes of the stereoscopically evoked components suggest that fewer neurons are activated synchronously by this stimulus, while the topographical differences prove that differently located or oriented neural generator populations are activated by three-dimensional stimuli.

These results are in line with single unit recordings in the visual cortex of monkeys, where depth-sensitive neurons were found (Poggio and Fischer 1977). Hubel and Wiesel (1970a) and von der Heydt et al. (1981) claimed that area 18 (which in man lies anterior to the primary visual area 17) contains more neurons responding to disparate stimuli. In single unit studies, sampling biases may mimic a differential distribution of functional units but evoked components reflect the mass activity of neural populations and global functional differences may be found. A more anteriorly located potential peak evoked by RDS stimuli in man, is a strong indication that stereoscopic stimuli activate more neurons in area 18 than in the primary visual cortex. The potential components evoked by contrast changes had the same latency (and a similar but nonidentical topography) as the stereoscopic components which are definitively of cortical origin. These findings thus strongly suggest that the P100 component evoked by two-dimensional contrast stimuli is of cortical origin.

2.3 Temporal Activity Patterns

Activity patterns of longer duration can be investigated topographically by plotting the most distinct features of the scalp fields (e. g., the locations of the potential maxima and minima) as a function of time. This yields information on the similarity of neural activation cycles evoked in different experimental conditions. The topographical differences between stereoscopic and contrast stimuli described in Sect. 2.2 are not restricted to the occurrence times of components: Fig. 9 shows that the differences prevail over long durations of time. The location means of 15 subjects are plotted at 5-ms intervals, and significant differences between stereoscopic and contrast evoked activity were established by computing paired t-tests as an exploratory statistic. The data of Fig. 9 suggest that the time course of the activation patterns in the two stimulus conditions is similar but that their topography is different. Both potential maxima and minima have different locations over longer epochs, indicating sustained differences in the activation of nonidentical neural generator populations.

Such neurophysiological differences are in line with functional differences between electrical brain activity evoked by dynamic RDS stimuli or by stimuli with

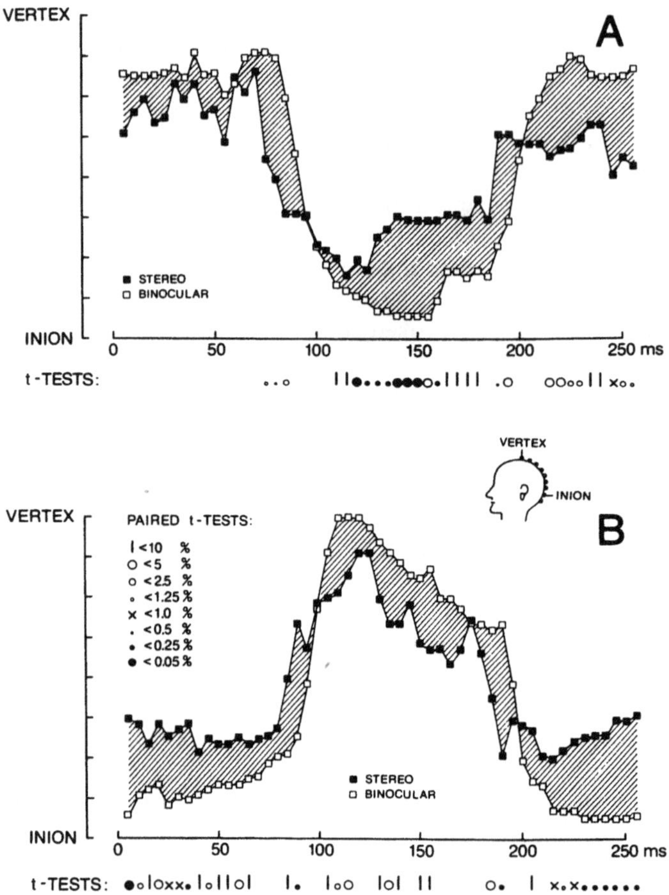

Fig. 9A, B. Mean location of potential maxima (**A**) and minima (**B**) of 15 subjects between 5 and 255 ms latency at intervals of 5 ms evoked by stereoscopic stimuli or by similar binocular contrast stimuli. *Hatched areas* indicate differences in location. The results of paired t-tests computed on the individual data are shown below each graph. Significant differences persist over longer time epochs. Note clusters of high significance levels

contrast borders. With temporally modulated patterns, Skrandies (1986b) showed that stereoscopic stimuli appear to persist visually at much lower time thresholds than contrast stimuli. In addition, with dynamic RDS stimuli a significant inverse linear relationship between temporal modulation frequency and evoked potential amplitude was found which was not observed with comparable contrast stimuli (Skrandies 1986b; Skrandies, in preparation).

2.4 Conclusion

The topographical analysis strategies outlined above are transparent and simple to apply, and they yield direct quantitative measures of component latencies and scalp locations. The derived measures are reference-independent and can be used for statistical comparisons between subject populations or between experimental conditions. Application of these topographical methods is not limited to potential fields; they have also been used successfully with multivariate statistical analysis: principal component loadings (Skrandies and Lehmann 1982b) and principal component scores (Skrandies 1983; Skrandies et al. 1984) have been topographically examined. The combination of multivariate statistics and scalp field distribution data proved a powerful tool in the analysis of differences in VEP lateralization (Skrandies and Lehmann 1982b) and of electrical brain activity recorded during the performance of a visual information-processing task (Skrandies 1983; Skrandies et al. 1984).

Other reference-independent descriptors of potential fields use information on the local gradients or on the curvature of the field distribution. The method of current source density analysis (Mitzdorf and Singer 1978; Mitzdorf 1985; McKay 1984a, b) or the "source derivation" as it became known in EEG analysis (Hjorth 1975), examines the second derivative of the potential distribution with respect to space. The magnitude of the local curvature of the potential field is estimated at each instant and can be plotted as a function of time, similar to the convential potential waveshapes. Such measures were displayed as chronotopograms by Rémond et al. (1969) who illustrated the curvature of spontaneous alpha EEG potential fields. Some of our evoked potential data have also been analyzed by computing the current source density. As expected, this method did not yield conclusions basically different from those that had been derived from direct spatial analysis (see Sects. 3.2.1 and 3.2.2, Figs. 13, 14 and 21).

3 Differences in Latency and Topography of Visual Evoked Potential Components

In an earlier study using multichannel recordings of visual evoked brain activity, Lehmann and Skrandies (1979) found significant differences in component latency between the potential fields evoked by upper or lower hemiretinal stimuli with upper retinal stimuli yielding shorter latencies than stimulation of the corresponding lower retinal areas. In order to generalize and to extend these earlier findings on checkerboard reversal stimuli (Lehmann and Skrandies 1979) in this section I will report supporting multichannel data. Four different series of experiments were performed: Potentials evoked by checkerboard reversal stimuli were recorded in 45 channels and were topographically analyzed as complete scalp field maps (Sect. 3.1). Additional experiments using pattern reversal stimuli are described in Sect. 3.2.1. Findings on potential fields evoked by different presentation modes of an identical checkerboard stimulus will be reported in Sect. 3.2.2, and the results obtained with grating stimuli will be discussed in terms of component latency and scalp topography in Sect. 3.2.3. In all the experiments, consistent differences were found in scalp topography and in component latency between the electrical brain activity evoked by upper vs lower retinal stimuli. These data suggest that there are basic functional differences between the upper and the lower hemiretinal systems.

In a further study the practical implications of these findings for the clinical use of visual evoked potentials were investigated (Sect. 3.3). One application of evoked potential recording in the visual modality relies on the exact determination of component latencies in patients suspected to have multiple sclerosis (Halliday et al. 1973). Prolongation of component latencies by only a few milliseconds may classify a patient's result as pathological, and our data obtained in a population of unselected neurological patients show impressively how the physiological differences between the two hemiretinal systems may cause a false pathological diagnosis if retinal stimulus location is not taken into account.

Both latencies and scalp locations of the evoked components show differences between upper and lower hemiretinal stimulation. Component latencies reflect functional characteristics of the hemiretinal system, while the scalp location differences are caused by the different cortical regions which are activated according to the retinotopic representation of the visual field in the visual cortex. Upper retinal stimuli are routed to regions above the calcarine fissure, while lower retinal stimuli are transmitted to cortical areas below the calcarine fissure (see Sect. 6.1.2). Topographically, this results in a more posterior scalp location for components evoked by lower hemiretinal stimuli projected to the same horizontal eccentricity. In many subjects, these occipital components are located well below the inion, and due to this fact, in some subjects the components are missed (Lehmann and Skrandies 1979; Lesèvre and Joseph 1979). Scalp electrode arrays extending to below the inion are thus mandatory for the recording of electrical brain activity evoked by lower retinal stimuli.

3.1 Complete Scalp Field Distributions: Pattern Reversal Evoked Activity Recorded in 45 Channels

Among 25 subjects studied, Lehmann and Skrandies (1979) reported on multichannel recordings (45 channels) from five subjects using a checkerboard reversal stimulus. In this section, additional 45-channel data evoked by upper or lower hemiretinal checkerboard reversal stimulation are presented.

Simultaneous recordings in 45 channels were obtained in six adult healthy volunteers. The electrode array extended to 70% of the distance from the inion to the nasion and to 3 cm below the inion, and it covered laterally 70% of the distance between the two meati acustici (see inset, Fig. 10). Technical details of the multichannel recordings can be found in Lehmann and Skrandies (1979).

A checkerboard reversal stimulus (50′ check size, 96% contrast) was presented on a circular field (15.5° arc diameter) via a rotating mirror at 1.6 reversals per second monocularly to the upper or lower hemiretina. The data were topographically analyzed following the principles outlined in Sect. 2.

Figure 10 illustrates two map series of scalp potential fields evoked by upper or lower hemiretinal stimuli. Equipotential maps are constructed by entering the potential value of each electrode at its respective position, and points of equal potentials are connected by lines (as contour lines connect locations of equal altitude on geographical maps). Relative global field power curves are also plotted for both map series. Maximal field power occurs for upper retinal stimuli at 113 ms and for lower retinal stimuli at 121 ms (see arrowheads). The two map series show a grossly similar pattern of activity at different post-stimulus times. An occipital positivity surrounded by steep gradients develops around 100 ms,

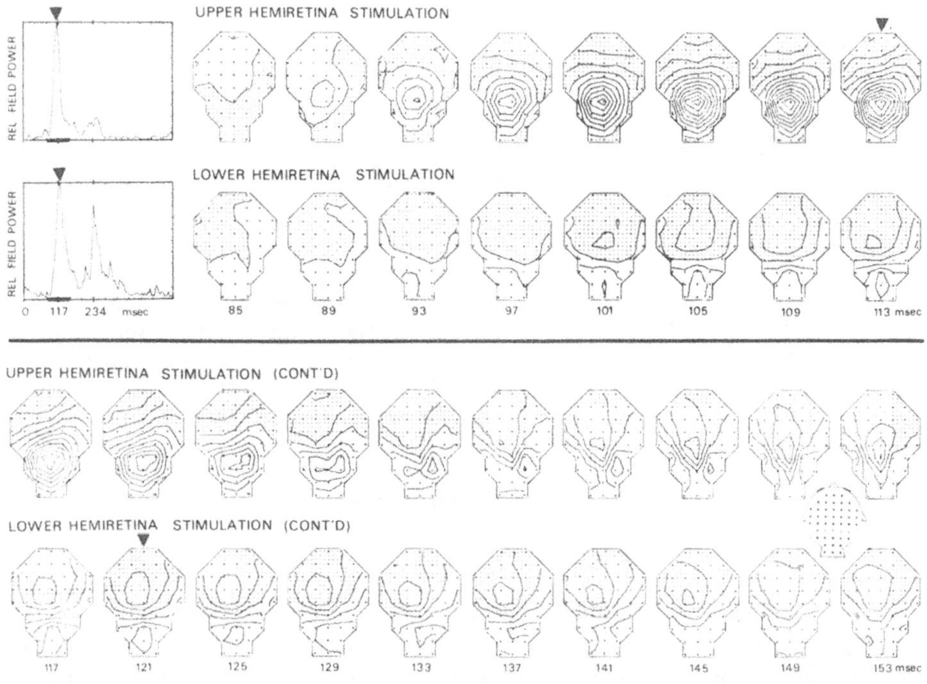

Fig. 10. Global field power curves (*top left*) and series of scalp field maps of potentials evoked by a checkerboard reversal stimulus (50′ check size, test field of 15.5° arc diameter, 96% contrast) presented to the upper or lower hemiretina. The 68-ms epoch during which maps are illustrated is indicated by *horizontal bars* in the field power plots. Simultaneous recordings in 45 channels as indicated in inset, most anterior electrode at 30% nasion-inion distance, circled electrode at inion. *Dotted areas* are negative, *white areas* positive with respect to the average reference. Equipotential lines in steps of 1 μV. *Arrowheads* in the field power graphs and over maps indicate times of maximal field power. Note the earlier latency time, and the more anterior scalp location of the occipital positive component for upper than lower hemiretinal stimuli. (From Lehmann and Skrandies 1984: by permission of Pergamon Press Oxford)

reaches its peak at component latency, and slowly disappears. This behavior is paralleled in time in the field power curves. There are also differences in the topographical distributions of the occipitally positive components, which show a more anterior scalp location of the components evoked by upper than by lower retinal stimuli.

The mean component latency of the six subjects was 99 ms for upper retinal stimuli and 115 ms for lower retinal stimuli. Longer component latencies for the lower retina were consistently found in all subjects, and the difference was statistically significant.

The scalp locations of the occipital positive components at individual times of maximal field power of the six subjects in the two stimulus conditions are illustrated in Fig. 11. In all subjects, components evoked by upper retinal stimuli are more anteriorly located than components evoked by lower retinal stimuli, resulting in a highly significant difference.

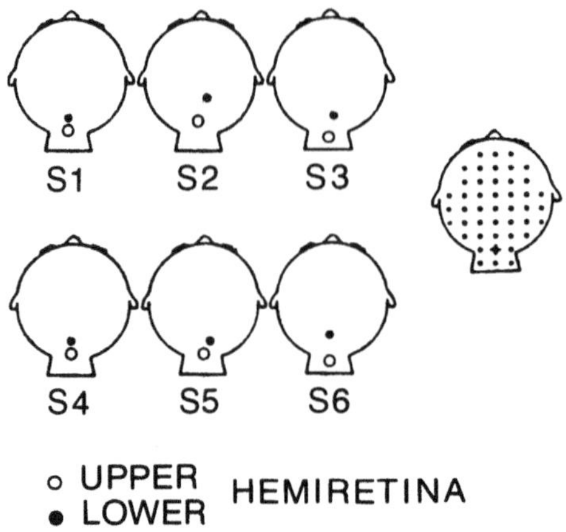

Fig. 11. Scalp locations of potential field maxima at individual times of maximal field power around 100 ms latency of six subjects evoked by upper retinal (*dots*) and lower retinal (*open circles*) stimuli. Mean component latency for the upper retina was 99 ms, for the lower retina 115 ms. Recordings from an array of 45 electrodes as illustrated. (Data from Skrandies and Lehmann 1982)

3.2 Generalization and Extension of Visual Evoked Potential Findings

3.2.1 Pattern Reversal Stimuli

Differences between potential components evoked by checkerboard reversal stimuli presented to the upper or lower hemiretina were further investigated in ten other subjects using a midline row of electrodes over the posterior scalp areas. Such a reduced electrode array was chosen because the potential fields obtained with 45 channels show occipital components which are symmetric about the midline (see Fig. 10). Thus, a row of electrodes along the sagittal midline detects the major pertinent features of the scalp field distributions.

Nine recording electrodes were placed at regular intervals of 7% of the nasion - inion distance along the midline between the vertex and the inion extending to 7% below the inion. Checkerboard reversal stimuli of 62% contrast (check size 30′ arc) were presented at a rate of 3.9 changes per second on a TV screen (8.7° by 11.4° arc) which was rotated by 90° so that the display lines ran vertically and significant display time differences were avoided.

Occurrence times of maximal field power between 80 and 140 ms determined component latencies. Component locations at these times and the topographical behavior of potential maxima and minima over the whole recording epoch were analyzed.

Figure 12 shows the grand mean potential waveshapes computed from the data of all subjects evoked by upper or lower hemiretinal stimuli. The time of maximal field power occurs later for the lower than the upper hemiretina (see vertical lines). Analysis of the individual field power data of the subject population resulted in a mean component latency of 110.8 ms for upper and 118.1 ms for lower retinal stimuli, a significant difference ($t = 4.45$, $p < 0.0025$).

Differences in topography may become more obvious in waveshapes when looking at the local curvature of the potential fields (see Sect. 2.4). Figure 13 illustrates how current source density data are obtained from the original data as the second derivative of the potential distribution with respect to space (Mitzdorf and Singer 1978; Mitzdorf 1985; McKay 1984a, b). The waveshapes display the change of the curvature of the field as a function of time, and thus the local characteristics of

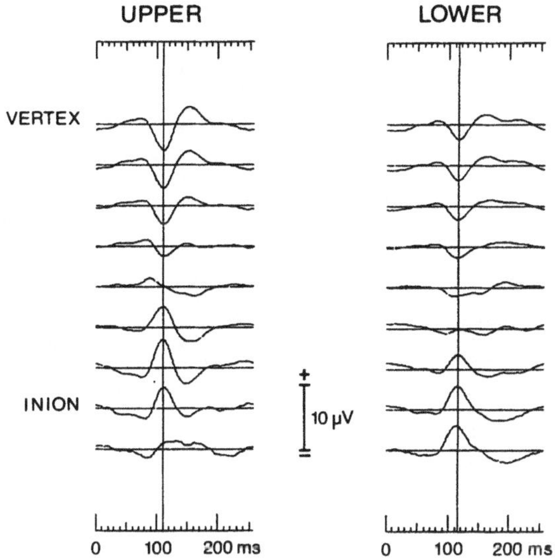

Fig. 12. Grand mean potential waveshapes of ten subjects evoked by upper (*left*) or lower (*right*) checkerboard reversal stimuli. *Vertical lines*, times of maximal field power. The electrode array starts at the vertex (Cz) and extends along the sagittal midline to below the inion. The data are referred to the average reference

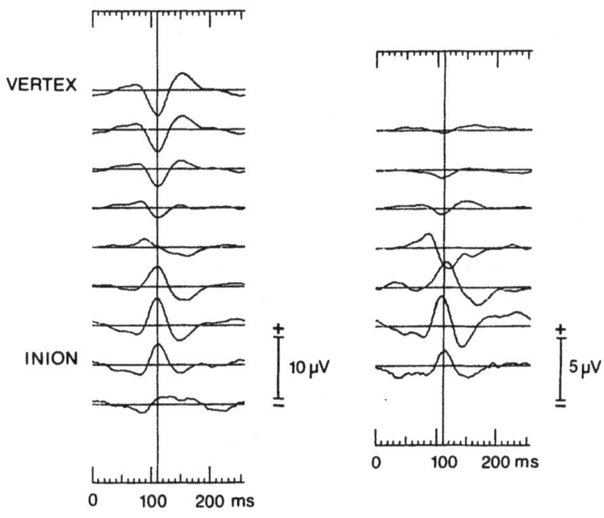

Fig. 13. Grand mean data (ten subjects) of potentials evoked by upper retinal stimuli. Potential waveshapes referred to the average reference (*left*) and corresponding current source density waveforms (*right*) illustrating the local curvature of the potential fields. Note that potential gradients become more pronounced in the current source density waveforms

the potential fields are enhanced. The grand mean data evoked by upper and lower retinal stimuli are compared in the form of current source densities in Fig. 14: While for the lower hemiretina the occipital positivity around 100 ms is restricted to the last two electrode positions, the upper hemiretina evokes a positive component extending further anteriorly. At electrode 5 a positivity is seen following upper hemiretinal stimuli, and a negativity following lower hemiretinal stimuli. The potential gradient is by definition reference-independent, and it is important to note that the direct analysis of the scalp fields in terms of field power and component location yielded identical topographical conclusions from the data.

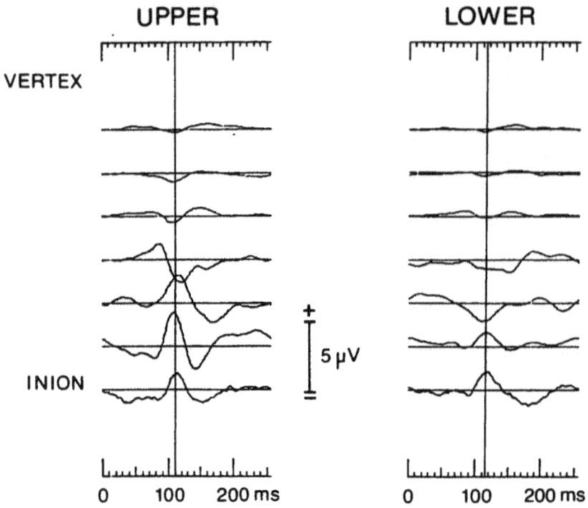

Fig. 14. Grand mean current source density waveforms of ten subjects evoked by upper or lower retinal stimuli illustrating the change of the curvature of the field as a function of time. This method enhances the local characteristics of the potential fields. The corresponding potential waveforms are shown in Fig. 13

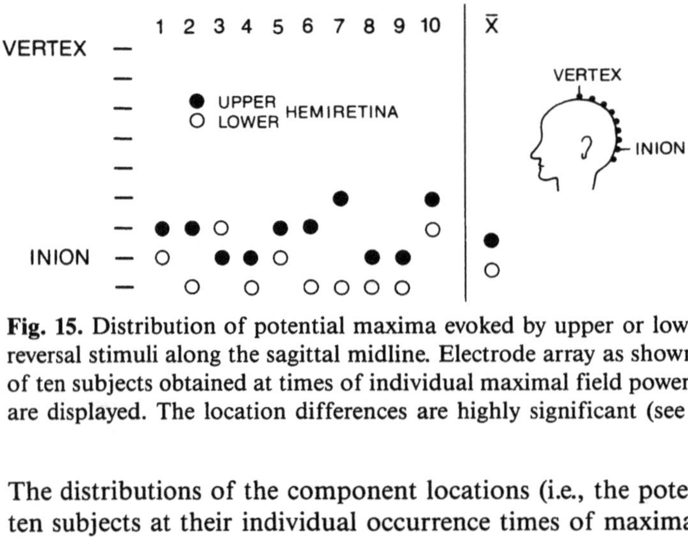

Fig. 15. Distribution of potential maxima evoked by upper or lower retinal checkerboard reversal stimuli along the sagittal midline. Electrode array as shown in head schema. Data of ten subjects obtained at times of individual maximal field power and population means are displayed. The location differences are highly significant (see text)

The distributions of the component locations (i.e., the potential maxima) of the ten subjects at their individual occurrence times of maximal global field power, as well as the mean component locations, are shown in Fig. 15. In nine of the ten subjects with upper hemiretinal stimuli, component locations are obtained which are anterior to those evoked by stimulation of the lower hemiretina, and the location differences were highly significant in the subject population ($t = 3.88$, $p < 0.0025$).

The differences of potential maxima and minima locations evoked by upper or lower retinal stimuli were further analyzed over the whole recording epoch of 256 ms. It turned out that location differences were not restricted to component latencies, but that the differences appear to be related to the complete neural activation cycle.

Figure 16A illustrates the mean potential maxima locations and indicates significance levels of the location differences between upper and lower retinal stimuli computed on the individual location data of the ten subjects. The mean locations of the potential minima and the corresponding significance levels are shown in Fig. 16B.

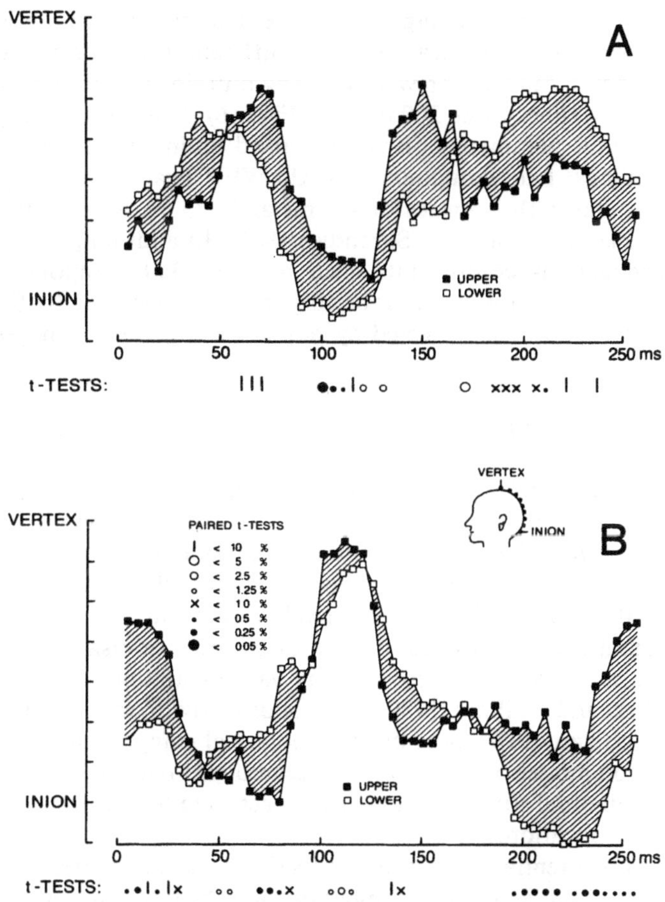

Fig. 16A, B. Mean potential maxima (**A**) and minima locations (**B**) of ten subjects evoked by upper or by lower retinal stimuli plotted between 5 and 255 ms at intervals of 5 ms. *Hatched areas* indicate differences in location. The results of paired t-tests which were computed on the individual data are shown below each graph. Significant differences persist over longer time epochs. Note clusters of high significance levels

The general distribution pattern for the maxima and minima is complementary in the anterior - posterior direction: during the analysis epoch the maxima exhibit up to about 80 ms an anterior, around 100 ms a posterior, and after about 130 ms again a more anterior scalp location, while the locations of the minima display a posterior - anterior - posterior sequence. This topographical pattern is very similar for brain activity evoked by upper and by lower retinal stimuli.

Most locations of the potential maxima evoked by upper or lower retinal stimuli are significantly different around 100 ms, and also around 200 ms (Fig. 16A). Significant differences of the minima locations are found at early times (up to about 30 ms) and around 100 ms, and high significance levels also cluster between about 190 and 255 ms (Fig. 16B).

The very consistent results suggest that different neural generating mechanisms are activated by upper and lower retinal stimuli: The brain electrical activation patterns corresponding to upper retinal stimulation are located more anteriorly on the scalp. Significant location differences are not only observed at component latencies, but they prevail over long durations. Differences are still found at long latency times after 200 ms (see Fig. 16). The long-lasting influence of physical stimulus parameters on visual evoked brain activity is in agreement with earlier findings. Late components of the VEP related to the spatial frequency of the stimulus with latencies of 188 ms and 335 ms have been reported for lateralized checkerboard stimuli (Skrandies 1981). In addition, the scalp topography of late endogenous components such as the P300 component, which is related to cognitive information processing, as first described by Sutton et al. (1965), was also found to be affected by retinal stimulus location (Skrandies 1983).

3.2.2 Different Presentation Modes: Onset, Offset, and Reversal

Patterned visual stimuli have been used successfully for the study of evoked potentials in man. VEPs elicited by contrast stimuli presented without gross luminance changes are more reliable and show less intersubject variation than flash VEPs, and have thus become standard for clinical purposes (e.g., Halliday et al. 1973). Differences in waveshape morphology evoked by pattern appearance (onset), disappearance (offset), or contrast reversal of an identical checkerboard pattern have been described by various authors (Estevez and Spekreijse 1974; Kriss and Halliday 1980). In a pilot study, Skrandies et al. (1980) demonstrated significant differences between potential components evoked by different presentation modes of an identical checkerboard pattern, and between upper and lower hemiretinal stimuli. These differences were found both in component latency and scalp location.

Scalp potential distributions of VEPs evoked by pattern onset, offset, and reversal stimuli presented to the upper or lower hemiretina were further investigated in 13 subjects using a midline row of nine recording electrodes.

In this section, it will be shown that different presentation modes of an identical checkerboard pattern evoke occipitally positive components which occur at different latencies (around 100 ms), and show scalp distribution differences suggesting the activation of non-identical neuronal populations. Simple waveshapes similarity between the reversal response and the offset response (Estevez and Spekreijse 1974; Kriss and Halliday 1980) does not appear to be sufficient evidence that these electrophysiological measures reflect identical underlying neuronal generating mechanisms, since different presentation modes give rise to neural mass responses which show distinct differences in time and scalp distribution, as will be shown below. In addition, these data demonstrate that the upper and lower retina behave as independent systems, displaying differences in component latency and in scalp location also with other stimulus presentation modes (Skrandies 1984a).

Thirteen healthy volunteers between 19 and 29 years of age (mean = median = 22 years) with normal or corrected-to-normal vision participated in the experiments. Nine Grass gold-cup electrodes were placed at regular intervals of 7% of the nasion - inion distance along the midline, starting at the vertex (Cz) and extending to 7% below the inion. The most anterior electrode served as recording reference.

The EEG was amplified (bandpassed from 0.1 to 70 Hz), A/D-converted at a rate of 1000 samples per second and channel over a period of 512 ms, averaged ($n = 120$), and stored on disc by a PDP11–40 computer.
Checkerboard reversal stimuli of 62% contrast were presented at a rate of 3.9 changes per second on a TV monitor (8.7° by 11.4° arc). Check size was either 60′ arc or 30′ arc. In the onset and offset conditions, the checkerboard was replaced every 256 ms by a homogeneous field of the same average luminance of 13.6 candela/m^2. When comparing upper and lower retina stimulation, significant display time differences inherent to a TV system were avoided by rotating the TV screen by 90° so that the display lines ran vertically. For hemiretinal stimulation, the subject fixated small marks at the upper or lower border of the screen from a distance of 130 cm.
The 12 experimental conditions (onset/offset/reverse * upper/lower hemiretina * large/small checks) were presented in a quasi-randomized order. Onset and offset responses were obtained in the same run. Each condition was presented three times; reliability of the data was confirmed by computing correlation coefficients between the first and the last run of the same experimental condition of each recording session using evoked potential waveshapes recorded at each electrode separately. Corresponding data of one subject participating in the experiment twice at a 1-week interval were also cross-correlated in order to determine the stability of the results.
The occurrence time of the maximal potential range between 80 and 160 ms coincided in most cases with the maximum of the first spatial derivative (i.e., the potential gradient) and defined component latency. In some subjects the onset responses were dominated by long latency components (as described by Kriss and Halliday 1980) or the lower hemiretinal responses displayed low signal-to-noise ratios, and in these cases (about 15%) the secondary maximal values of the potential range function were included as an additional criterion. Comparison of this strategy with the unrestricted component definition did not reveal any differences regarding the conclusions derived from the data.
For statistical analyses, mean values of component latency and location were derived from repeated measurements in each subject and were entered in separate three-factorial analyses of variance. Significant influences were further examined by computing t-tests using adjusted alpha levels according to Holm's (1979) procedure. All results presented were significant at an alpha level of $p < 0.005$.

3.2.2.1 Reliability of the Data

The mean correlation coefficients over all experimental conditions estimating the within-session reliability ranged between 0.60 and 0.90, with a mean value over all subjects of 0.75. The replication of the experiment in one subject at an interval of 1 week yielded a mean correlation of $r = 0.83$ (range 0.74–0.93). Since all 256 time points were used for computing the correlations, not only major components but also very early and late parts of the VEP contribute in the same way to the general waveshape similarity indexed by the correlation coefficients. Thus, the values reported above indicate both a high overall similarity of the repeated recordings and reliability of the data.

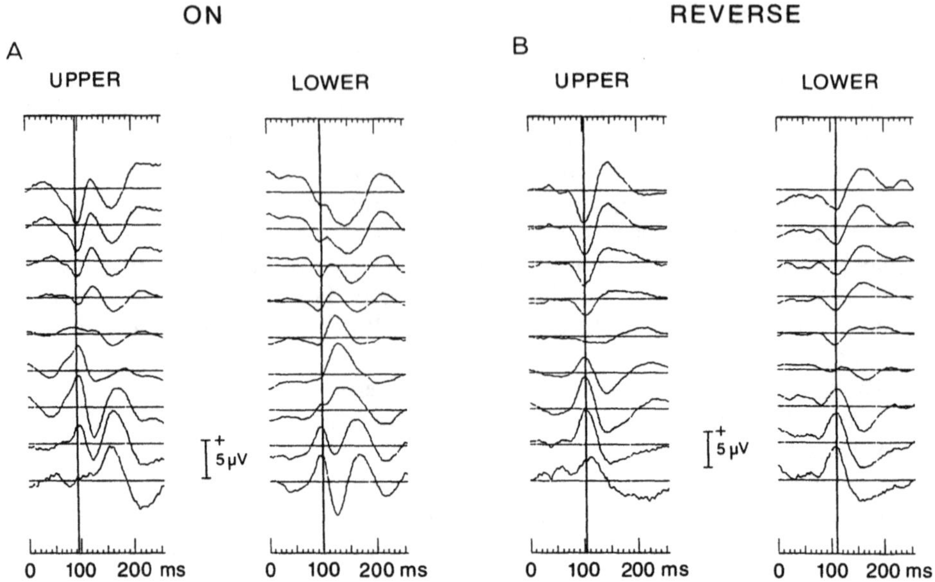

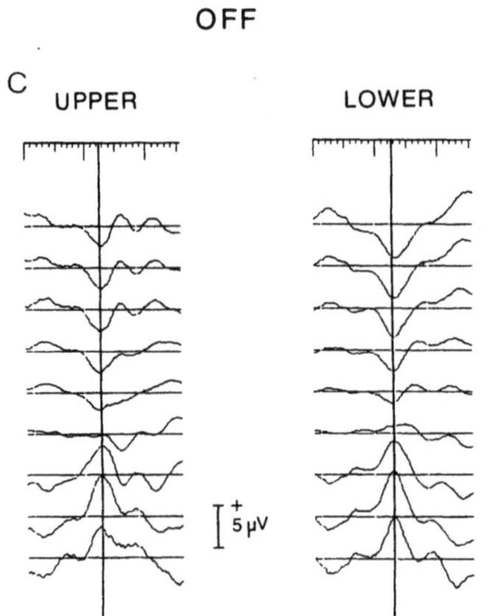

Fig. 17A–C. Original potential waveshapes evoked by small checks presented to the upper or lower retina. Recordings are obtained from a row of nine midline electrodes between the vertex (first trace) and 7% below the inion. A checkerboard pattern (check size 30′ arc) was presented as onset (**A**), reversal (**B**), or offset stimulus (**C**). Average reference data from one subject

3.2.2.2 Differences of Component Latencies

Figure 17 illustrates sets of original VEP waveshapes of one subject evoked by a checkerboard presented to the upper or lower hemiretina as onset (Fig. 17A), reversal (Fig. 17B) or offset stimulus (Fig. 17C). The major component in all experimental conditions is an occipital positivity occurring around 100 ms latency, and there appear systematic differences in component latency, as illustrated by the

Table 1. Mean component latencies (ms) of 13 subjects and latency differences between stimulus conditions

Stimulus	On	Difference	Reverse	Difference	Off
Upper hemiretina	94.2	13.9***	108.1	15.9***	124.0
Difference	16.5***		8.6***		10.4***
Lower hemiretina	110.7	6.0**	116.7	17.7***	134.4

***$p<1\%$; **$p<5\%$: t-tests corrected according to Holm's procedure

potential range plots in the corresponding Figs. 18A–C. The vertical lines in Fig. 17 indicate times of maximal field power, and one sees that the component latencies are different for different presentation modes as well as for upper and lower retinal stimuli.

The occipital positive component evoked by pattern onset showed consistently shorter latencies (mean over all subjects: 102.5 ms) than that evoked by reversal (mean: 112.4 ms), or offset stimuli (mean: 129.2 ms). Three-factorial analyses of variance showed significant effects of presentation mode and of retinal location on component latencies. Check size was not a significant factor and also did not show up in interactions. Thus, the mean component latencies evoked by large and small stimuli were used for further analysis. Table 1 summarizes the latency data and indicates significant differences revealed by paired t-tests with alpha levels corrected for multiple testing according to Holm's (1979) procedure. Onset stimuli yielded significantly shorter latencies than reversal or offset stimuli for both the upper and lower hemiretina, while offset stimuli consistently showed the longest latencies.

Significant latency differences were also found when comparing component latencies evoked by upper or lower retinal stimulation: Upper hemiretinal stimuli yielded shorter latencies than lower hemiretinal stimuli. This finding was significant for all three presentation modes (see Table 1).

The finding that onset stimuli yielded significantly shorter latencies than offset stimuli is supported by other data on motor reaction time. Pease and Sticht (1965) reported that the reaction times to stimulus onset were significantly shorter than those to stimulus offset when the light stimuli were presented to the peripheral retina. In the fovea no statistically significant reaction time differences were observed. This observation was confirmed by Bartlett et al. (1968), who found significantly shorter reaction times to onset than to offset stimuli. These differences were independent of the wavelength of the stimulus.

3.2.2.3 Differences of Component Locations

Topographical differences between conditions were revealed by comparison of the data as potential profiles on the scalp plotted over time (see Sect. 2.1.2). Figure 18 illustrates the data of Fig. 17 in the form of series of topographical potential profiles. In each profile series one sees that during the first 60 ms a flat potential distribution prevails, and then an occipital positivity develops which reaches its

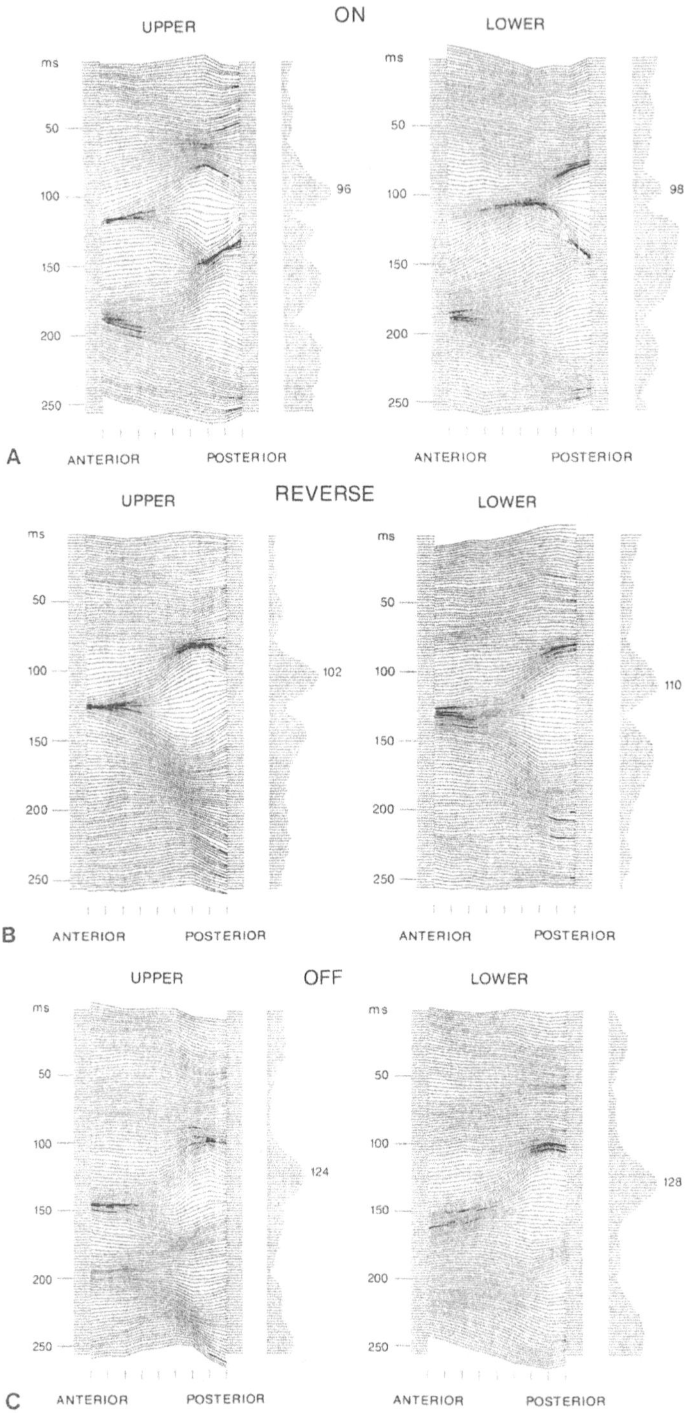

Fig. 18A–C. Data of Fig. 17 displayed topographically as series of potential profiles. Adjacent to each profile series, global field power is plotted. Different presentation modes yield topographical differences. **A** Onset; **B** reversal; **C** offset stimuli

maximum between 96 and 128 ms (depending on the stimulus condition). Later this maximum is replaced by a potential minimum. This behavior over time is similar to that of the complete scalp field map series illustrated in Fig. 10. The amount of activity quantified by the global field power at each instant is plotted as a function of time adjacent to each profile series in Fig. 18.

Although the potential profiles in Fig. 18A–C are globally similar, there are distinct differences: Not only the latencies but also the respective component locations appear to depend on presentation mode and retinal stimulus position. Three-factorial analyses of variance yielded significant effects of presentation mode and retinal location as well as a significant interaction of these factors. This indicates that different presentation modes do not display the same location differences between components evoked by stimulation of the upper or lower hemiretina. As with the latency data, check size turned out not to be a significant factor, and mean values of the location data evoked by large and small checks were used for comparisons with paired t-tests. Again, alpha levels were corrected for multiple comparisons (Holm 1979).

The results on component location are summarized in Table 2. For the upper hemiretina all component locations were significantly different from each other depending on presentation mode: pattern onset components were located most anteriorly (mean 5.14 cm above the inion) and offset components (mean location at the inion). On the other hand, components evoked by lower hemiretinal stimulation displayed no significant location differences according to presentation mode.

Highly significant differences were observed for onset and reversal stimuli when comparing retinal location: The components were located more posteriorly following lower hemiretinal stimulation. Components evoked by offset stimuli showed no significant differences between upper and lower hemiretinal stimuli. This may be explained by the fact that the components evoked by upper offset stimuli are already located far posteriorly on the scalp, and with lower retinal stimuli no recordable differences are present.

Figure 19 shows the grand mean potential profiles of the 13 subjects in all stimulus conditions which were investigated. The profiles were constructed at the individual occurrence times of the components and then averaged across subjects. The topographical differences discussed above are also evident in the grand mean data.

Table 2. Mean component locations of 13 subjects and location differences between stimulus conditions

Mean peak position (cm above inion)

Stimulus	On	Difference	Reverse	Difference	Off
Upper hemiretina	5.14	3.09***	2.05	2.05***	0.00
Difference	4.63***		2.56***		0.77
Lower hemiretina	0.51	1.02	−0.51	1.28	0.77

***$p < 1\%$: t-tests corrected according to Holm's procedure

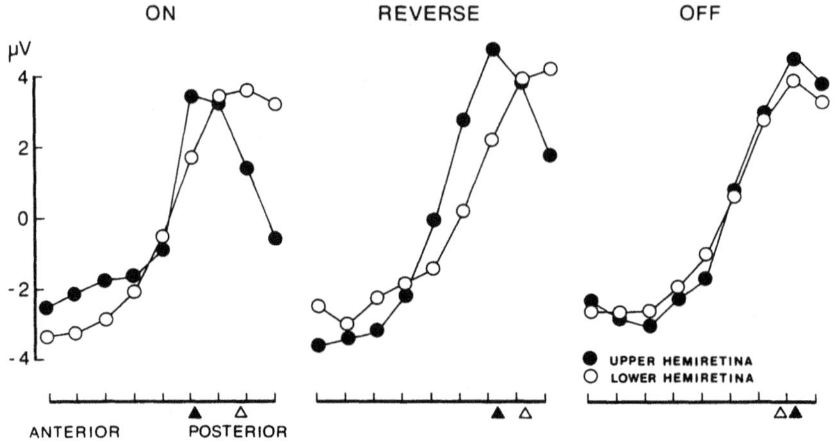

Fig. 19. Grand mean potential profiles of 13 subjects in the various stimulus conditions at individual times of maximal field power. Note location differences between upper and lower retinal stimuli as well as differences between presentation modes. *Arrowheads* indicate mean component locations

Additional paired t-tests confirmed that there were no significant differences between conditions in terms of potential range, amplitude, or gradient. The global appearance of the scalp distributions is rather similar, showing an occipital positive component around 100 ms latency. Small but statistically significant changes occur mainly in the location of the maximal scalp field values (arrowheads in Fig. 19, see also Table 2). This suggests that similar neural generating mechanisms taking place at different post-stimulus times are activated within different regions of the occipital cortex.

3.2.2.4 Analysis of Waveshape Similarity

As the detailed topographical analysis shows, visual potentials evoked by different presentation modes of an identical checkerboard pattern yield different neural activation patterns. These differences are both in the time domain (i.e., component latencies) and in the spatial distribution of the evoked electrical fields. None of the presentation modes showed sufficient similarity to support the identity of the potentials evoked by offset and reversal that has been suggested (Estevez and Spekreijse 1974; Kriss and Halliday 1980). Also, a simple summation of onset and offset responses does not yield the same pattern of activity as that evoked by reversal stimuli. Figure 20 illustrates this point: In all recording channels the computed sum of onset and offset response waveforms looks different from the potentials evoked by a checkerboard reversal stimulus. The question of waveshape similarity was approached quantitatively by computing cross-correlation functions between corresponding channels evoked in the different stimulus conditions. Table 3 shows the numerical results, giving the mean and the median correlation coefficients as well as the maximal correlation obtained in any one of the nine recording channels. The data in the upper right triangle of the matrix of Table 3

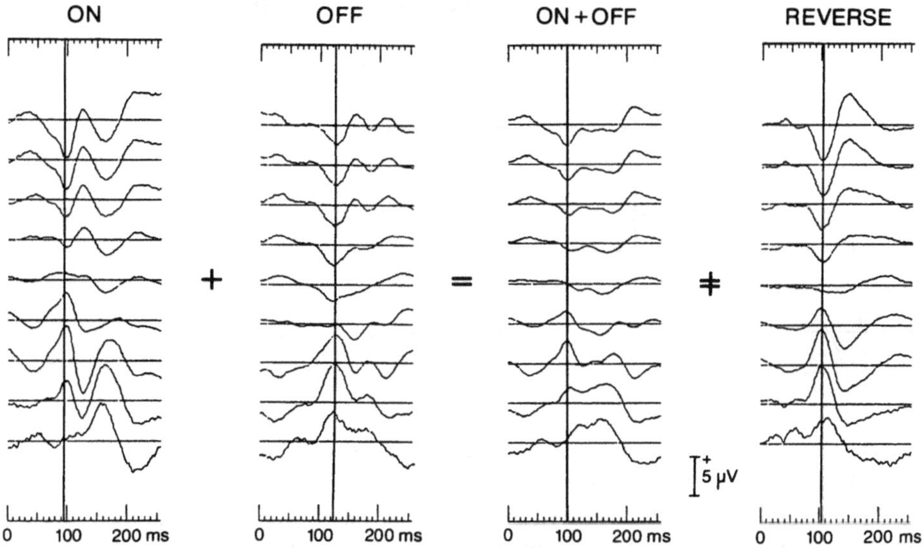

Fig. 20. Potential waveshapes evoked by stimuli presented to the upper retina as onset, offset, or reversal presentations. The computed mean of onset and offset evoked potentials (*ON+OFF*) is different from the reversal potential waveshapes

Table 3. Correlation coefficients computed between potential waveshapes from different stimulus conditions for corresponding recording channels

Maximum	0 ms	On	Off	Reversal	On + off
On	Mean Median Maximum		0.05 −0.01 0.58	0.06 0.07 0.30	0.77 0.79 0.91
Off	Mean Median Maximum	0.52 0.55 0.65		0.29 0.22 0.63	0.64 0.60 0.87
Reversal	Mean Median Maximum	0.66 0.68 0.84	0.66 0.66 0.76		0.25 0.17 0.76
On + off	Mean Median Maximum	0.78 0.79 0.91	0.71 0.79 0.88	0.72 0.75 0.79	

Mean and median values are derived from the correlations in all nine channels, maxima give the largest correlation coefficients obtained independent of the channel. Upper right triangle lists correlation coefficients at time zero (i.e., no time shift allowed). Lower left triangle gives correlation coefficients derived from maxima of whole cross correlation functions

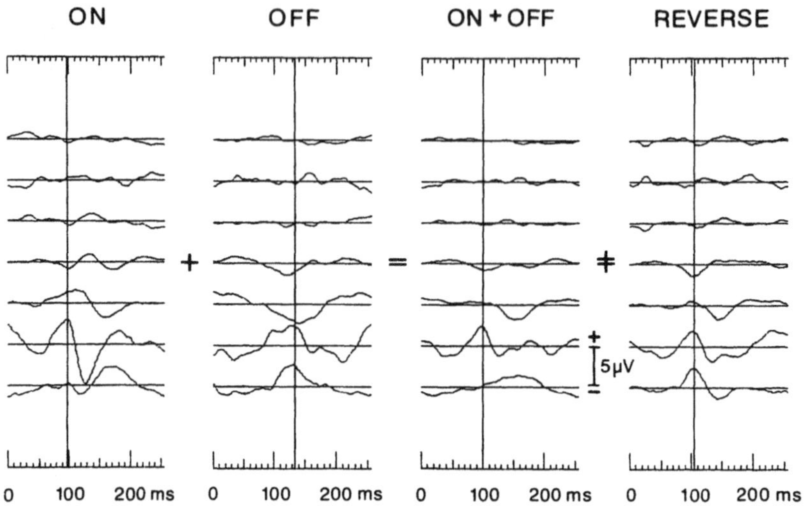

Fig. 21. Current source density waveforms evoked by onset, offset, and contrast reversal stimuli illustrating the change of the curvature of the potential field as a function of time. The computed sum of onset and offset evoked brain activity is also included. Note pronounced differences in latency and in local potential field curvature between stimulus conditions. The corresponding potential waveshapes are shown in Fig. 20

correspond to the correlations obtained at time zero (i.e., when not considering any time shifts present), and one sees that both onset and offset evoked potentials are more similar to the computed sum (the mean of onset and offset responses) than the reversal condition, which yields a mean correlation of only $r = 0.25$. When the maxima of the cross-correlation functions are analyzed (i.e., when the time series are shifted against each other, and only the maximal correlation coefficient is considered), basically the same result is obtained. These correlation coefficients, given in the lower left triangle of the matrix of Table 3, are higher than the other values. This is to be expected, since the curves have been shifted with respect to each other in order to maximize similarity. But still the correlations between the computed sum of onset and offset responses and the reversal responses are smaller than the correlation between onset or offset responses and the computed mean waveshapes. As a result of the correlation analysis it is clear that the potential waveshapes also show differences, and that there is no compelling similarity between any of the potentials evoked by the different presentation modes.

In addition, the current source density of these data was computed, resulting in the waveshapes shown in Fig. 21. This figure illustrates the second spatial derivative, which by definition is independent of the reference electrode, and it is obvious that stimulus onset, offset, and reversal evoked spatially different electrical brain activity patterns. The latency differences described above are confirmed (see vertical lines in Fig. 21), and the largest changes in the curvature of the evoked scalp fields occur in the onset condition further anterior on the scalp than with checkerboard offset or reversal. Comparing the activity of channel 5, it becomes obvious that reversal and offset are also different. These data further support the conclusion that electrical brain activity elicited by contrast reversal is not identical to a simple summation of the brain activity evoked by stimulus appearance and disappearance.

3.2.3 Grating Stimuli

In another population of 11 adult subjects, recordings in 16 channels to a reversing vertical grating stimulus presented to the upper or lower hemiretina were obtained. Scalp field distributions evoked by stimuli of different spatial frequencies show differences in component latency and potential range, and such multichannel data are discussed in detail elsewhere (Skrandies 1984b). In the present experiments it was found that also with grating stimuli, the evoked scalp field distributions of all subjects display differences both in component latencies and in topography, depending on retinal stimulus location.

Evoked potentials were recorded in 11 healthy subjects (between 19 and 42 years of age) from an electrode array consisting of 16 electrodes and extending in four regular steps from 2 cm below the inion to 7 cm above the inion, and in two steps to 6 cm left and right from the midline (see inset in Fig. 22). A frontal midline electrode served as reference for recording. For amplification, Grass P511 amplifiers with a bandpass between 0.1 and 100 Hz were used, and the data were A/D-converted at a rate of 1000 samples per second and channel over a period of 256 ms, averaged ($n = 90$), and stored on disc for later analysis.

Square-wave-modulated grating stimuli of 80% contrast were presented at a rate of two reversals per second by a TV stimulation system as a circular test field of 7° arc diameter. Vertical gratings of 2.3 cycles per degree were presented to the upper or lower retina from a distance of 185 cm.

Potential field maps were constructed from the data, and global field power defined the component latencies. Differences of component latencies and locations were tested using paired t-tests.

Figure 22 shows the grand mean potential waveshapes evoked by upper hemiretinal stimuli. Vertical lines indicate times of maximal field power (see also Fig. 26). In the lower part the construction of the potential field map at component latency is illustrated. A series of potential distribution maps is illustrated in Fig. 23 in which the data of Fig. 22 are shown between 80 and 138 ms with intervals of 2 ms.

Grand mean potential waveshapes computed from the data of all subjects evoked by lower hemiretinal stimuli are illustrated in Fig. 24. The scalp distribution at component latency shows a potential maximum shifted posteriorly as compared to the activity evoked by upper retinal stimuli. The whole series of scalp field maps is different from that evoked by upper hemiretinal stimuli in terms of component location (see Fig. 25). Latency differences determined by global field power are shown in Fig. 26: The upper hemiretina reaches its maximum at 97 ms, the lower hemiretina at 99 ms. These figures refer to the grand mean data (i.e., the mean evoked activity of the 11 subjects studied), and paired t-tests on the individual data revealed consistent, statistically significant differences. The mean latency for the upper hemiretina was 96.4 ms, while that for the lower hemiretina was 100.7 ms in the subject population. This difference was significant ($t = 4.66$, $p < 0.0005$).

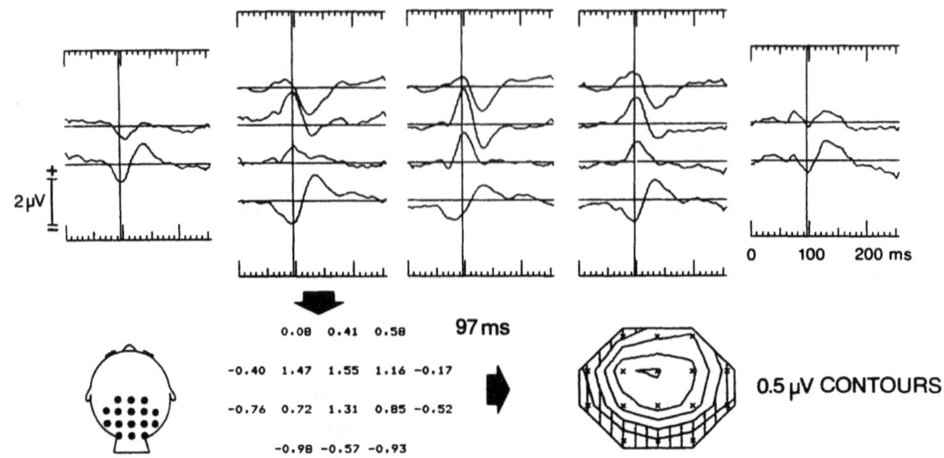

Fig. 22. Grand mean evoked potential waveshapes of 11 subjects evoked by upper hemiretinal stimulation, recorded from an array of 16 electrodes extending in four regular steps from 2 cm below the inion to 7 cm above the inion, and in two steps to 6 cm to the left and right of the midline (see inset). At component latency (maximal field power at 99 ms) the construction of a potential map is illustrated: the microvolt values of each electrode are entered at their respective positions (*lower left*), and areas of equal potential values are connected by lines. This results in an equipotential line map (*lower right*). (From Skrandies 1986a: by permission of Butterworth Publishers, Stoneham)

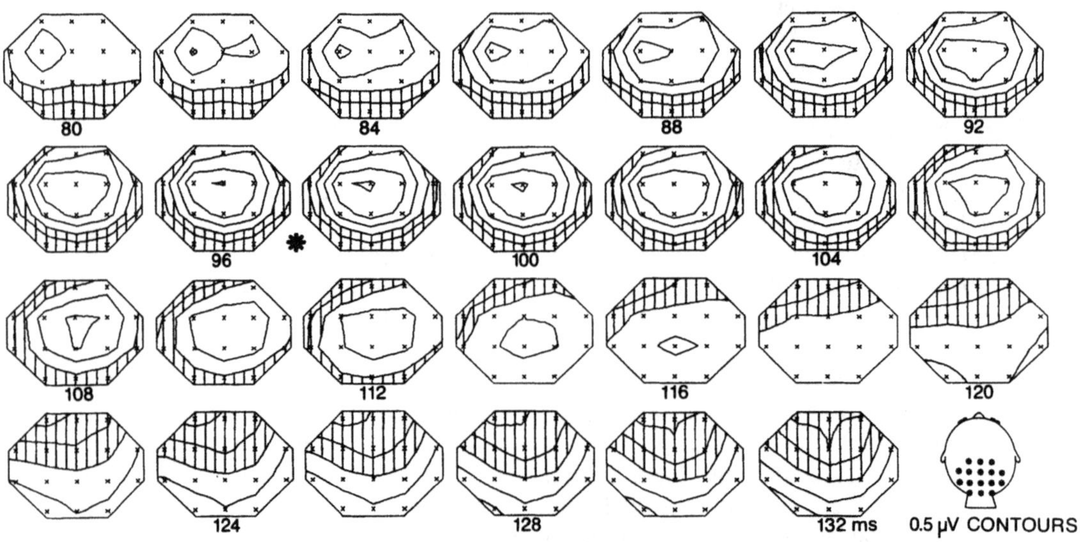

Fig. 23. Series of grand mean potential field maps evoked by upper hemiretinal stimuli between 80 and 132 ms at 2 ms-intervals (*hatched areas* negative with respect to the average reference). *Asterisk* indicates time of maximal global field power. Same data as Fig. 22. (From Skrandies 1986a: by permission of Butterworth Publishers, Stoneham)

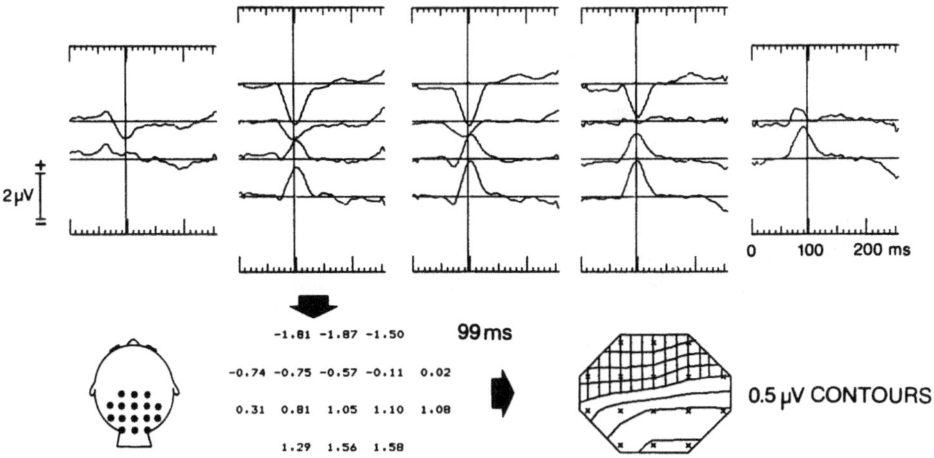

Fig. 24. Grand mean evoked potential waveshapes of 11 subjects evoked by lower hemiretinal stimulation. As in Fig. 22, the construction of a potential map at component latency is illustrated. (From Skrandies 1986a: by permission of Butterworth Publishers, Stoneham)

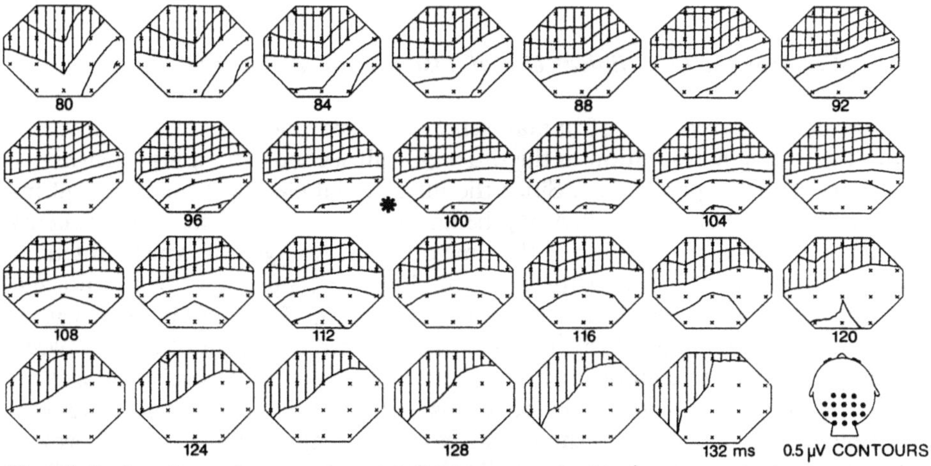

Fig. 25. Series of grand mean potential field maps evoked by lower hemiretinal stimuli between 80 and 132 ms at 2 ms-intervals (*hatched areas* negative with respect to the average reference). *Asterisk* indicates time of maximal global field power. Same data as Fig. 24. (From Skrandies 1986a: by permission of Butterworth Publishers, Stoneham)

Fig. 26. Global field power of grand mean data of Figs. 23 and 24 in 11 subjects. Field power values are scaled to 100%. Note that the maximum occurs earlier for upper than lower hemiretinal stimuli (see *arrowheads*)

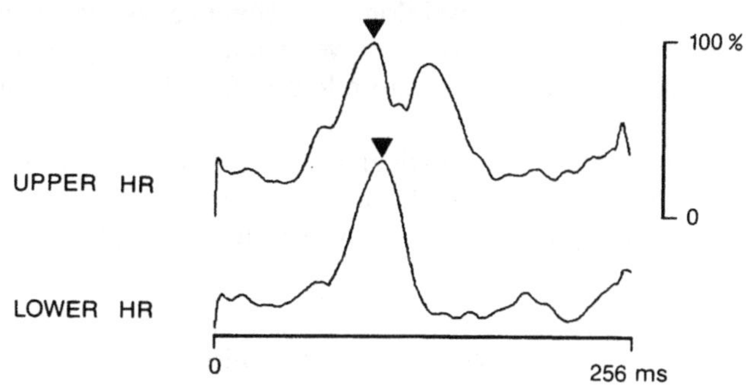

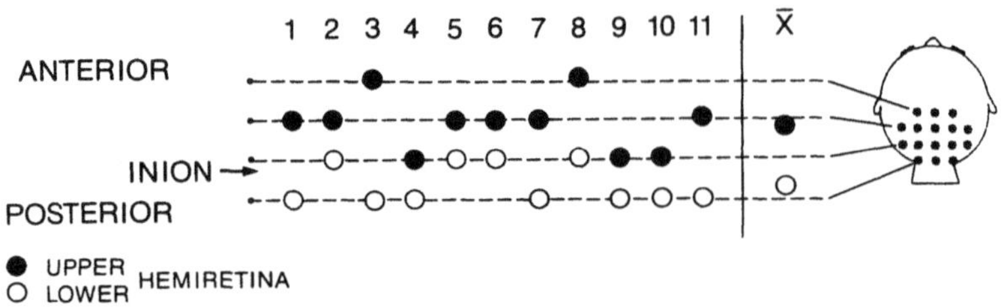

Fig. 27. Distribution of potential maxima in the anterior-posterior direction of 11 subjects at individual component latencies for upper and lower hemiretinal grating stimuli, and mean locations (*right*). The location differences are statistically significant ($t = 7.45$, $p < 0.0005$). Individual data of Figs. 23 and 25 were used. Electrode positions as indicated on the *right*

Component location differences were also tested using the individual data as shown in Fig. 27, where component locations evoked by upper of lower hemiretinal stimuli are illustrated in the anterior - posterior direction. The mean location difference was 4.62 cm ($t = 7.45$, $p < 0.0005$).

3.3 Implications for the Clinical Use of Visual Evoked Potentials [2]

The conventional recording of human sensory evoked potentials is a noninvasive method with a time resolution in the order of milliseconds and a high sensitivity to sensory stimulus parameters. Since the intracranial electrical fields spread instantaneously via volume conduction, the latency of evoked potential components gives unambiguous information in the time domain. Analysis of evoked potential latency is considered a useful measure for diagnostic purposes. In the normal population, the most replicable VEP component is the P100 component, with a latency of approximately 100 ms and a small standard deviation of only between 3 and 6 ms. (Of course, both mean and standard deviation depend on the exact physical stimulus parameters.) Owing to this low variability, latency measures can be employed for detecting pathological cases. A major application of evoked potential recording in the visual modality is the establishment of clinically silent damage to the optic nerve in patients suspected of multiple sclerosis (Halliday et al. 1973). Latency prolongation of only a few milliseconds may identify a patient's result as pathological. However, such latency differences may also be caused by the existing physiological differences between the upper and lower retinal systems, as I have shown above. Therefore, it appeared of interest to investigate whether the differences are relevant for diagnostic conclusions in a clinical setting.

325 unselected neurological patients suspected to have multiple sclerosis (i.e., "possible" and "probable" cases according to the scale of McAlpine et al. 1972) were investigated. The only criteria for inclusion in the study were visual acuity

[2] Patient data were obtained from routine examinations at the Neurology Department of the University Hospital, Zürich. The permission to use the data is gratefully acknowledged

above 0.8 and reliable evoked potential recordings. This yielded VEP data from 512 eyes of 325 patients.

Evoked potentials were recorded from an array of four electrodes equidistantly spaced in the midline between 2 cm below and 7 cm above the inion referred to a frontal electrode at Fz. Checkerboard reversal stimuli of 96% contrast (60′ arc check size) were presented as a circular target of 15.5° arc diameter at a distance of 90 cm to the upper or lower hemiretina. Repeated runs were obtained of each condition, and component latency was determined by the occurrence time of the maximal potential range within the recording array.

Patients were classified as pathological when one or both eyes had component latencies exceeding the normal range, defined as latency values above the mean plus three standard deviations of a group of healthy controls investigated earlier (Lehmann and Mir 1976).

The patients were classified according to various combinations of upper and/or lower retinal component latencies. The numbers and percentages of pathological and normal cases depended on whether only upper, only lower, both upper and lower, or either upper or lower evoked potential latencies were considered:

1. When either upper or lower retinal evoked potential latencies were taken as criterion of normality, 257 of the patients (79.07%) had to be classified as pathological.
2. Only 68 patients (20.92%) showed normal latencies for both upper and lower retinal stimulation.
3. In all cases where the upper hemiretinal stimuli showed prolonged latencies, the lower hemiretinal components were also abnormal.
4. Following upper retinal stimulation, in only 69 patients (21.23%) did the components evoked exceed the normal latency range, and 256 patients (78.76%) had normal latencies for both eyes. These data will be further analyzed in detail below (see Fig. 28).
5. In 188 (73.44%) of the 256 patients with normal upper hemiretinal latencies, the lower hemiretinal evoked component latencies were pathological, while the classification of 68 patients (26.56%) remained normal.

Figure 28 shows the frequency distribution of component latencies in the two stimulus conditions of the 256 patients for whom normal latencies were obtained in both eyes with upper retinal stimulation. The latency values of the lower retina are shifted toward longer latencies, and they have a much broader distribution than upper hemiretinal latencies. The upper retina yielded component latencies with a mean of 100.5 ms and a standard deviation of 5.2 ms, whereas components evoked by the lower retina had a mean of 118.1 ms and a standard deviation of 14.5 ms. Paired t-tests confirmed that the latency differences were highly significant ($p<0.0001$).

In addition, upper retinal stimuli resulted in normally distributed latencies, while the latency values of the lower retinal stimuli appeared to deviate from a Gaussian distribution (Fig. 28). The distribution parameters skewness and kurtosis were computed and found to differ for the two stimulus conditions. The upper retinal data were not significantly different from a normal distribution (skewness -0.163, $z=1.51$, n.s.; kurtosis -0.257, $z=1.19$, n.s.) whereas the lower retinal data deviated from normality (skewness 1.259, $z=11.66$, $p<0.0001$; kurtosis

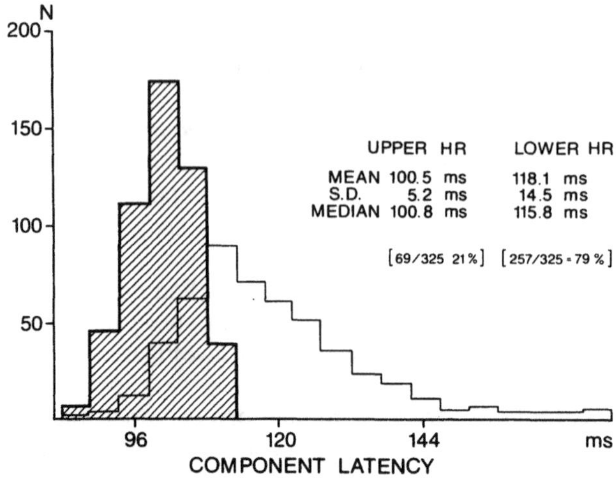

Fig. 28. Frequency distribution of component latencies of 256 patients in which normal latencies were found for upper retinal stimulation of both eyes. The latency values of the lower hemiretina (*thin line*) show a broader distribution than upper hemiretinal latencies (*thick line*). Mean, standard deviation, and median values of the population as indicated. The data obtained for the upper retina are not significantly different from a normal distribution, while the lower retina data deviate from normality (see test)

2.621, $z = 12.16$, $p < 0.0001$). These differences were tested by the procedure described by van der Waerden (1971, pp. 229–230).

In summary, the data of a population of unselected neurological patients show impressively how inherent physiological differences between the two hemiretinal systems increases the variability of evoked potential measures. The high intersubject variation of component latencies evoked by the lower hemiretina suggests that this stimulus condition is not at all appropriate for diagnostic purposes: Because of the large standard deviation, which is nearly three times that of the upper hemiretina, the normal range becomes very large. When lower instead of upper retinal stimuli were used, the discriminative power between normal and pathological results decreased and the number of false-negative increased (i.e., pathological cases become classified as normal).

More importantly, the physiological differences as reflected by different component latencies may cause a false pathological diagnosis if retinal stimulus location is not taken into account. A simple change of fixation toward the lower parts of the stimulus display is followed by prolonged component latencies. Typically, the stimulus displays used are not much larger than 20° and "wrong" fixations may easily occur. In our data the number of pathological latencies increases from 69 to 257 patients (i.e., by 57.84%) when the lower retinal data are used. For clinical purposes it appears of major importance to carefully control the patients' fixations. The upper hemiretinal responses yield more stable results with less intersubject variability and thus prove more useful for meaningful diagnostic interpretations.

4 Electrophysiological Differences at Other Stages of the Visual System

In the preceding section evidence was obtained that stimuli presented to the upper hemiretinal system are processed more efficiently than those presented to the lower hemiretinal system. The difference is reflected in shorter latencies of the cortical evoked components. In this section I will investigate whether such differences can already be detected at more peripheral parts of the visual system. The electro-oculogram (EOG) can be used as an electrophysiological method to study changes of the intraocular standing potential of the eye which shows a long latency response to the onset of steady illumination. This potential is believed to reflect the rate of metabolism elicited by light stimuli (Arden and Kelsey 1962b; Marmor and Lurie 1979), and the use of small localized test lights may reveal local differences at the first step of visual information processing (Sect. 4.1).
In electroretinography (ERG) contrast alternating patterns can be used as localized retinal stimuli which elicit the so-called pattern ERG. The pattern ERG is believed to originate from retinal structures associated with the ganglion cell layer, and the experiments presented below will demonstrate differences between upper and lower retinal areas reflected in the human pattern ERG (Sect. 4.2).

4.1 Retinal Pigment Epithelium Activity [3]

Eye movements can be electrically recorded from skin electrodes placed at the canthi of the eye because the vertebrate eye forms an electrical dipole with the cornea positive with respect to the fundus. This is the electrophysiological basis of eye movement recordings (EOG). The EOG can be used as an indirect method to study changes of the intraocular potential when eye movements of a constant amplitude are made (Arden and Kelsey 1962a, b). The short-latency a-, b-, and c-waves of the ERG [in the order of milliseconds, or seconds in case of the c-wave (Granit 1947; Armington 1974)] are elicited by the onset of steady illumination of the eye, and a large positive deflection, called the "light peak," occurs only after several minutes. Investigations of human patients (Arden and Kelsey 1962b; Täumer 1976), show that the retinal pigment epithelium is necessary to produce the light peak, which is interpreted as an indicator of the rate of metabolism (Marmor and Lurie 1979). Valeton and van Norren (1982), Linsenmeyer and Steinberg (1982), and Griff and Steinberg (1982) performed intraocular microelectrode recordings in various species, and their data suggest that the source of the light peak is located in the basal membranes of the retinal pigment epithelium. The standing potential of the eye is recorded in man indirectly by the EOG, and its covariation with stimulus parameters allows conclusions to be drawn about the functional interaction between the pigment epithelium and the retinal photoreceptor layer under different experimental conditions.
In the experiments presented in this section, the EOG was used in order to investigate whether electrophysiological differences between the upper and lower retinal areas are restricted to evoked potential measures (see Sect. 3), or whether differences can be detected already at a very basic retinal level. The activity of the

[3] Results of this section have been published elsewhere (Skrandies and Baier 1986)

pigment epithelium as quantified by EOG measurements of perifoveal upper and lower retinal regions will be described. Absolute differences in response to a small localized light stimulus, as well as the different reactivity of upper and lower retinal areas to changes of the test light intensity, will be reported.

Twelve healthy subjects between 19 and 32 years of age participated in this study. The measurements extended over five different sessions lasting about 2 h each, and in order to minimize diurnal variations all experiments were performed at the same time of day (late in the afternoon). The pupils were dilated maximally by the application of Mydriaticum Roche® and Neosynephrine®, and Hellige Ag-AgCl recording electrodes were fastened at the lateral and medial canthi of each eye.

During an adaptation period of at least 60 min, the subject was exposed to a white adaptation light of -1.77 log candela/m^2 which was projected as a Ganzfeld onto a half-cylindric acrylic screen. After about 1 h the stability of the baseline was tested. The subject performed at 1-min intervals horizontal eye movements with an excursion of 40°. About 8 min later the standing potential (i.e., the EOG amplitude) had reached a steady state, and a circular white test light of 10° diameter with a luminance of 2.73 log candela/m^2 was back-projected at the center of the Ganzfeld. The adaptation light remained on throughout the session. Foveal, and peripheral upper and lower retinal areas at 10° and 40° eccentricity along the vertical meridian were stimulated in different experimental sessions in a quasi-randomized order by changing the location of fixation marks.

The EOG was recorded with a "Tönnies-EOG-System" consisting of a DC amplifier (low pass at 20 Hz), and an analog computer which integrated the amplitudes of ten consecutive eye movements. The raw EOGs and the integrated values were plotted online on an x-y plotter. A baseline was defined as the mean amplitude of the last eight recordings of the adaptation period.

In order to discriminate the light peak from spontaneous fluctuations of the dark value, the 95% confidence limit of the values obtained in the adaptation period was determined for each eye separately using the t-distribution. Peak values exceeding this confidence limit were considered reliable, and only two of the 120 measures obtained (12 subjects $\times 2$ eyes $\times 5$ retinal locations) had to be excluded. The light peak amplitudes were expressed as percentages of the adaptation values, and the means of the right and left eye of each subject were used for further analysis. Because percentage values are not normally distributed, Wilcoxon tests were used for comparing experimental conditions.

An additional series of experiments was performed with three subjects, each of whom participated in 15 sessions. In these experiments we investigated the reactivity of upper and lower retinal areas to different test light intensities. The same procedure as described above was used, but the measurements were restricted to upper and lower retinal areas at 10° eccentricity. The test light intensities varied between 2.42 and 3.63 log candela/m^2 in steps of 0.5 log candela/m^2. This yielded five light peak values for each location, and the relationship between EOG amplitudes and light intensity was established by computing linear regressions.

The original EOGs of one subject measured at five different retinal locations are illustrated in Fig. 29A. About 10 min after the light onset the light peak occurs in all recordings with a mean peak latency of 9.5 min in the subject population. The latencies were similar for all retinal locations, and there were also no significant differences between central and peripheral light peak latencies.

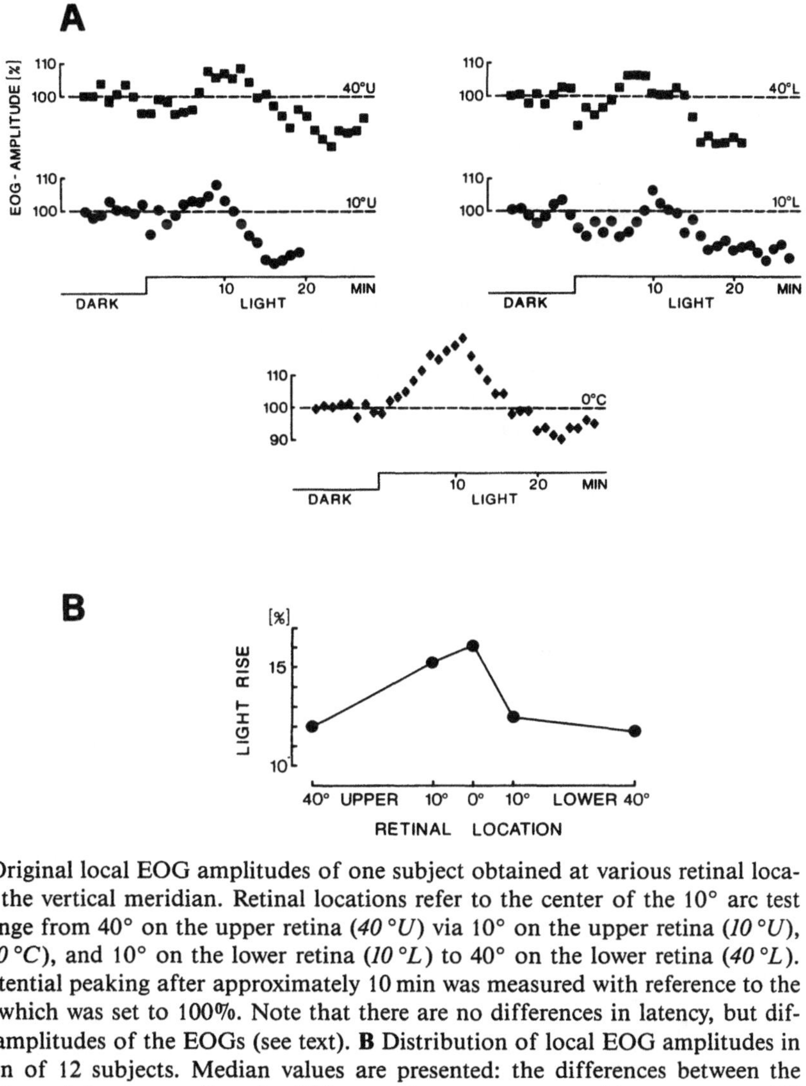

Fig. 29. A Original local EOG amplitudes of one subject obtained at various retinal locations along the vertical meridian. Retinal locations refer to the center of the 10° arc test light and range from 40° on the upper retina (*40 °U*) via 10° on the upper retina (*10 °U*), the center (*0 °C*), and 10° on the lower retina (*10 °L*) to 40° on the lower retina (*40 °L*). A rise in potential peaking after approximately 10 min was measured with reference to the base value, which was set to 100%. Note that there are no differences in latency, but differences in amplitudes of the EOGs (see text). **B** Distribution of local EOG amplitudes in a population of 12 subjects. Median values are presented: the differences between the center and the periphery, as well as the differences between upper and lower areas at 10° eccentricity, were statistically significant. (After Skrandies and Baier 1986)

The light peak amplitudes were in the order of only 10% – 20% above the baseline condition. This is accounted for by the small test light used. The highest amplitudes of the light peak occurred after central stimulation (median over all subjects: 16.12%), and central light peak amplitude increases were significantly larger than peripheral light peak amplitude increases (median difference 2.47%, $p<0.05$). Perifoveal upper retinal areas had higher amplitudes of the light peak than the corresponding lower retinal areas. However, these differences were significant only at 10° eccentricity, where upper retinal areas showed a median light peak amplitude rise of 15.26% while lower retinal areas had a median light peak of 12.44% (median difference 2.82%, $p<0.025$). At 40° eccentricity the same tendency was found (12.06% light peak amplitude for the upper, 11.76% for lower retinal areas), but due to the large intersubject variation the difference was not statistically significant (median difference 0.30%, n.s.). The difference between the results obtained at 10° and at 40° eccentricity may be explained by the fact that rod densities are much higher in perifoveal areas than farther out in the retinal periphery (Østerberg 1935). This may well account for the higher variability observed at 40° eccentricity.

The population data are summarized in Fig. 29B showing the distribution of the median light peak amplitudes at the five retinal locations examined. The upper retinal areas have higher amplitudes than the corresponding lower retinal areas, and central stimuli yield the highest amplitudes.

With increasing intensity of the test light the light peak amplitudes increased, and a linear relationship was observed in all subjects. The mean correlation coefficient between EOG amplitudes and the logarithm of the test light intensities was 0.75. Figure 30 illustrates the slopes of the regression lines computed individually for each subject for upper and for lower retinal stimuli. In order to enable a direct comparison between retinal locations independent of the absolute EOG amplitude differences, the regression lines were normalized, and Fig. 30 shows the light peak amplitudes relative to the amplitudes found with a test light intensity of 2.5 log candela/m^2. Upper retinal stimuli consistently elicit steeper slopes than lower retinal stimuli. This suggests a greater sensitivity of the upper retinal areas to changes in light intensity.

Light peak amplitude increases in the order of less than 20% were also reported in other EOG studies in which small localized test stimuli were used (Aschoff

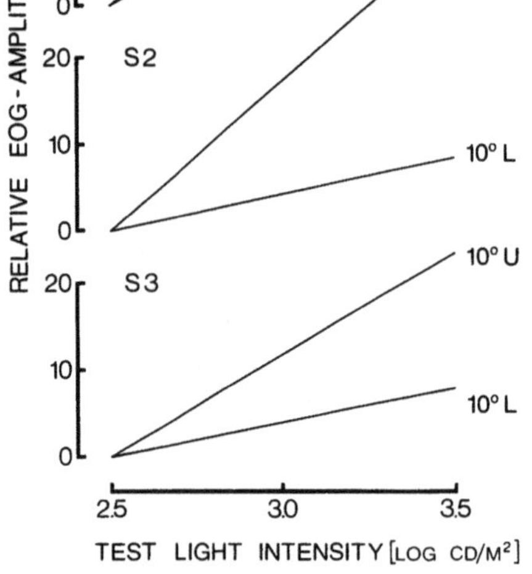

Fig. 30. Slopes of regression lines of three subjects indicating the relationship between relative EOG amplitude and the logarithm of the test light intensity. Stimuli presented 10° above the fovea (*10 °U*) yield steeper slopes than stimuli 10° below the fovea (*10 °L*). Data were obtained at five light intensities between 2.42 and 3.63 log candela/m^2 with intervals of 0.5 log candela/m^2. In order to enable a direct comparison between upper and lower retinal responses independent of the absolute EOG amplitudes, the values recorded at 2.5 log candela/m^2 were set to 0%. (After Skrandies and Baier 1986)

1981). The finding of slightly, but statistically significantly higher EOG light peak amplitudes for foveal than perifoveal stimuli, in the order of 2.5%, confirms reports on the contribution of cone activity to the light peak amplitude (Afanador and Andrews 1978). This interpretation is in line with data obtained from congenital achromats (Elenius and Aantaa 1973), and similar differences between EOG light peak amplitudes elicited by central or peripheral stimuli were observed by Aschoff (1981) and Dodt and Baier (1984). The differences between the upper and lower perifoveal retinal areas at 10° eccentricity were small but consistent and statistically significant in the subject population. At 40° eccentricity similar differences were observed but the variability of the EOG amplitudes was high, and the differences between upper and lower retinal stimuli did not achieve statistical significance. The smaller variability of the data at 10° eccentricity is explained by the fact that a larger number of photoreceptors are activated by a test light located between 5° and 15° eccentricity than by stimuli located between 35° and 45° eccentricity (see Østerberg 1935).

The systematic differences of light peak amplitudes cannot be caused by intrasubject variation in the various test sessions. In addition, variation of electrode placement or electrode impedance, which may cause different absolute EOG amplitudes, cannot account for the effects of retinal location in our data, because the light peak amplitudes were always referred to the EOG values obtained in the baseline condition at the end of the respective dark adaptation periods.

In addition to the absolute light peak amplitude differences there were also relative differences in amplitude change when the test light intensity was varied. As expected from the results of Arden and Kelsey (1962b), the light peak amplitude was proportional to the logarithm of the test light intensity. Upper retinal areas consistently showed a larger increase of the EOG response, which suggests that the upper retina is more sensitive to changes of light intensity.

From clinical studies it is known that the response of the standing potential to light requires contact between the photoreceptors and the pigment epithelium, and thus the light peak is believed to reflect the rate of metabolism elicited by light stimuli (Arden and Kelsey 1962b; Marmor and Lurie 1979). The light peak presumably originates in the basal membranes of the pigment epithelium (Valeton and van Norren 1982; Linsenmeyer and Steinberg 1982; Griff and Steinberg 1982). The finding that stimulation of upper retinal areas at 10° eccentricity is followed by significantly larger light peak amplitudes than stimulation of lower retinal areas suggests that stimuli presented to upper retinal areas elicit a higher level of light-induced activity, reflecting the interaction between the photoreceptors and the retinal pigment epithelium, than the stimulation of the corresponding lower retinal areas. Anatomical differences in the regional distribution of rods and cones in the human retina, where a higher density of photoreceptors is reported for upper retinal areas (Østerberg 1935), support this interpretation. The amplitude of the light peak can be interpreted as reflecting the rate of metabolism elicited by light stimuli, and thus should correlate with the anatomically determined photoreceptor density (see also Sect. 6.1.2).

In summary, the EOG data show that absolute and relative differences between upper and lower retinal areas in response to light stimuli can be found at the first step in visual information processing at a very basic retinal level.

4.2 Human Pattern Electroretinograms

Contrast alternating patterns produce no stray-light artifacts and have thus been used as localized retinal stimuli in ERG studies (Riggs et al. 1964; Armington 1968). Such stimuli elicit the so-called pattern ERG that can be recorded from the intact human eye. The origin of the components of the pattern ERG is different from that of those of the luminance ERG, which reflects the activity of the retinal layer of photoreceptors (a-wave) and the activity of bipolar and Müller cells (b-wave). While the exact generating mechanisms of the components of the pattern ERG remain a matter of discussion (cf Riemslag et al. 1985), this experimental technique can certainly be employed to investigate potentials generated at a higher retinal level than the pigment epithelium activity described in Sect. 4.1.

There is some evidence that the pattern ERG originates from retinal structures associated with the ganglion cell layer. Maffei and Fiorentini (1981) and Maffei (1982) have shown that the pattern ERG of the cat diminishes gradually after transection of the optic nerve and eventually disappears after about 4 months, while the luminance ERG remains unchanged. Holländer et al. (1984) confirmed these findings and correlated the attenuation of the pattern ERG responses with the retrograde changes in retinal morphology following the section of the optic nerve. These authors demonstrated that their electrophysiological findings were paralleled by morphological changes restricted to the nerve fiber and ganglion cell layers of the retina. Similar pattern ERG findings in the monkey after sectioning the optic nerve were reported by Maffei et al. (1985).

Clinical observations support this interpretation: Groneberg and Teping (1980) described pattern ERG recordings from a patient with a traumatic lesion of the optic nerve. Shortly after the lesion both pattern and luminance ERGs could be recorded, whereas about 3 months later the pattern ERG had disappeared while the luminance ERG remained unchanged.

The experiments presented in this section were aimed at demonstrating differences between upper and lower retinal areas, reflected in the human pattern ERG, which may be associated with the activity of retinal ganglion cells.

In order to arrive at meaningful conclusions from pattern ERG recordings in man, it appeared of major importance first to establish that the electrical activity recorded (i.e., the components of the pattern ERG) is of retinal origin. Surprising results were reported from experiments where pattern ERGs were recorded from an eye which was occluded while the contralateral eye was stimulated (Seiple and Siegel 1983). Following the considerations on the scalp topography of electrical fields outlined in Sect. 2, a reference-independent measurement and analysis strategy was applied to simultaneous recordings of evoked potentials and pattern ERGs. In Sect. 4.2.2 experiments will be reported in which significant differences between upper and lower retinal areas were found using pattern ERG recordings.

4.2.1 Spatial Distribution of Pattern Electroretinograms and Evoked Potentials

In order to prove that the pattern ERG has its origin in the eye and is not a far-field potential originating in the visual cortex, simultaneous multichannel recordings were made at the eye and at several scalp locations. Gold foil electrodes were

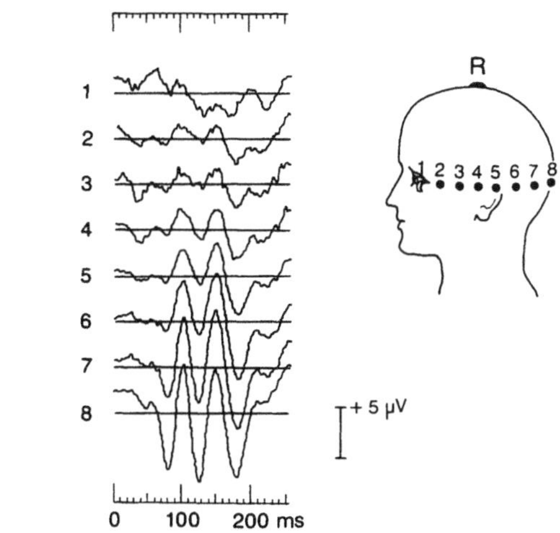

Fig. 31. Potential waveshapes recorded from a row of eight electrodes covering laterally the distance between the left eye and the inion (see head schema). Reference electrode (*R*) at the vertex (Cz). Checkerboard reversal stimuli evoked a pattern ERG around 60 ms in the eye channel and VEP activity around 100 ms in the occipital channels

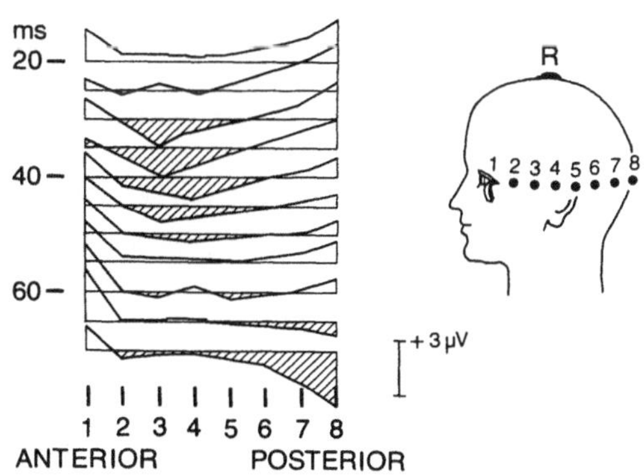

Fig. 32. Data of Fig. 31 shown as potential profiles along the head between 20 and 70 ms latency. Note the consistent shape of the anterior potential gradients

used to record the pattern ERG from the eye (Arden et al. 1979). Two different scalp electrode montages were used. In one experiment, eight recording electrodes were placed in a line along the side of the head from the left eye to the inion with a vertex reference (Fig. 31). In a second topographical experiment, eight recording electrodes were spaced equidistantly from the eye via the forehead along the midline of the head to the inion using the ear as reference point (Fig. 33). The reference sites were selected merely for convenience in order to enable waveform comparisons of the pattern ERG with the published literature; they had no consequence for the analysis of the electrical field distribution (Sect. 2).

A topographical distribution with large potential gradients around the eye is expected for retinal potentials, whereas visual cortical sources produce steep potential gradients over the occipital areas. Figure 31 shows the potential waveshapes recorded to a checkerboard reversal stimulus (check size 30′ arc, 80% contrast) fixated centrally. At early latency times, a positive deflection appears after about 60 ms in the first channel while the other recording channels show little activity. Around 100 ms on the other hand, waveform peaks occur in all recording channels, and the positivities are most pronounced over the occipital scalp areas. The topographical changes over time along this lateral row of electrodes are illustrated in Fig. 32 as a series of potential profiles. Between 40 and 70 ms a potential peak

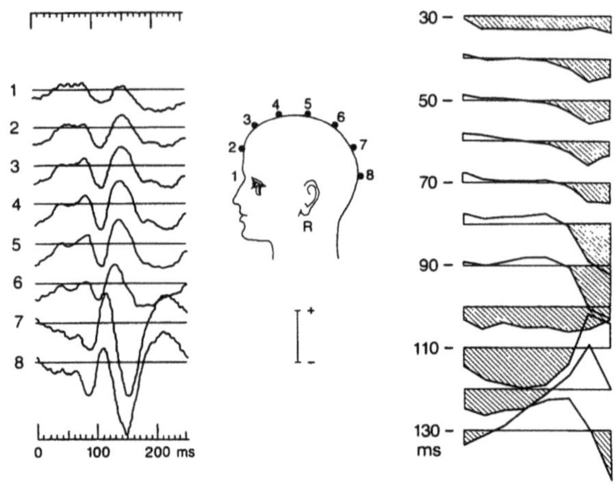

Fig. 33. Potential waveshapes and scalp potential distributions evoked by a checkerboard reversal stimulus presented to the right eye. The left eye was occluded, and activity was recorded in eight channels along the midline between the left eye and the inion using a reference electrode (*R*) at the ear. Channels 1 – 6 are on nearly equipotential levels (potential profiles on the *right*) while occipital recording channels show VEP activity after 100 ms. Calibration: 5 μV

builds up at the eye electrode, reaching its maximum at 65 ms. This early latency range is dominated by a steep potential gradient between the eye electrode and the second electrode, whereas the potential gradients between channels 2 and 6 are shallow. The gradients over the occiput (between channels 7 and 8) show less consistent behavior during these early latency times: between 20 and 60 ms the gradient is first positive, and then becomes negative. In contrast, the anterior potential gradient remains negative throughout the analysis period considered in Fig. 32. These differences between the anterior and the occipital recording channels suggest that the source of the evoked activity recorded in the first channel as a pattern ERG is located in the eye.

Similar conclusions arise from data obtained from a different subject with a midline electrode montage along the head from the left eye via the forehead to the inion. in these experiments the effect of occluding the recorded eye was also investigated. Seiple and Siegel (1983) had reported that a pattern ERG waveform can be recorded even when only the fellow eye is stimulated. We repeated this stimulus condition: Fig. 33 shows the data obtained when the right eye was presented a checkerboard reversal stimulus while the ERG activity was recorded from a row of eight midline electrodes equidistantly spaced between the left eye and the inion. The anterior electrodes of the array appear to be on nearly equipotential levels, whereas in the posterior recording channels the occipitally originating evoked potential activity is seen. The corresponding potential profiles are shown in Fig. 33 on the right. In the anterior channels, small potential gradients are seen throughout the analysis period. Occipitally, a relative negativity dominates the times around 80 ms, and at 110 ms the scalp distribution shows an occipital positivity surrounded by steep gradients. This scalp topography is similar to that illustrated before for components evoked by checkerboard reversal stimuli (see potential waveshapes and profile in Fig. 5, map series in Fig. 10, profile series in Fig. 18). From these topographical multichannel data it is obvious that the waveshape recorded at the eye cannot be interpreted as a pattern ERG. The scalp distribution evoked by the same checkerboard reversal stimulus presented to the left eye is illustrated in Fig. 34. The potential waveshapes are similar to those evoked by stimulation of the right eye except for the electrical ac-

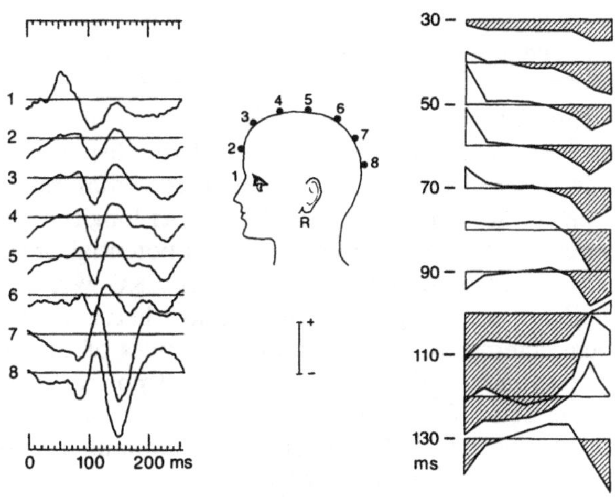

Fig. 34. Potential waveshapes and scalp potential distributions evoked by a checkerboard reversal stimulus presented to the left eye. Electrical activity was recorded in eight channels along the midline between the left eye and the inion. Channels 2 through 8 show a similar pattern of activity throughout the recording period, as in Fig. 33. The eye channel displays a pattern ERG around 50 ms resulting in steep anterior potential gradients. Same subject as in Fig. 33. Calibration: 5 μV

tivity recorded in the first channel, which is reminiscent of the pattern ERG. The potential profiles on the right of Fig. 34 around 50 ms have steep gradients anteriorly (i.e., between the eye electrode and electrode 2) while around 110 ms there is the same occipital positivity as in Fig. 33.

The topographical analyses of the pattern ERG and VEP show that retinal activity can be meaningfully interpreted and distinguished from contamination of cortical sources by considering electrical scalp field distributions. For examining upper and lower retinal areas, similar multichannel recordings were obtained, and the latency of the pattern ERG components was determined by the occurrence of a maximal potential gradient at electrodes at and near the eye. This method is independent of the reference electrode, and it avoids confusion generated by waveshape analysis (Seiple and Siegel 1983).

4.2.2 Differences Between Upper and Lower Retinal Areas

Simultaneous recordings of pattern ERGs and potentials evoked by central, upper, or lower retinal stimuli were obtained in a population of 11 healthy adult subjects.

VEPs were recorded from six electrodes spaced equidistantly along the midline between 40% of the nasion-inion distance above the inion and an electrode 10% below the inion. Pattern ERGs were recorded from the left eye using a gold foil electrode (Arden et al. 1979) inserted under the lower lid. Two skin electrodes were placed at the outer canthus of the eye and 3 cm further posterior. The distances between electrodes 1 and 2 and electrodes 2 and 3 were equal.

The EEG and ERG signals were amplified, bandpassed between 0.1 and 70 Hz, and A/D-converted at a rate of 1000 samples per second over a period of 256 ms. In order to avoid artifact contamination of the ERG by eye movements and blinks, 12 averages consisting of only 30–60 sweeps were obtained for each experimental condition in a quasi-randomized order. The data were edited offline, and for final analysis artifact-free grand means of each retinal location were computed for each subject individually.

Checkerboard patterns consisting of checks of 30′ arc and 62% contrast were reversed at a rate of 3.9 reversals per second on a TV monitor (rectangular screen of 8.7° by 11.4°) which was rotated by 90°. The subject's head was supported by a chin-forehead rest, and he looked binocularly at fixation marks in the center or at the upper or lower border of the display from a distance of 130 cm.
The vertex electrode served only as recording reference; for analysis of the VEP data the six scalp sites were treated as longitudinal potential profiles, while the pattern ERG data were analyzed as potential profiles along the three electrodes at the eye. Components of the pattern ERG were defined by the maximal potential gradient occurring in a time window between 35 and 75 ms. VEP components were determined by the occurrence times of the maximal potential range which corresponds to the multichannel field power measure between 80 and 140 ms.
Three sets of data of a typical subject are illustrated in Fig. 35. The first waveshape shows the pattern ERG, computed as the difference between the eye electrode and the electrode at the outer canthus; the other waveshapes are the scalp channel data which are computed vs the average reference. In all stimulus conditions the pattern ERG has a similar waveshape with a positive component occurring after about 50–60 ms. For central retinal stimuli the pattern ERG amplitude tends to be larger than for upper or lower retinal stimuli. The potential gradients within the three-channel recording array were larger for central stimuli than for peripheral stimuli in all but four of 22 cases.
As shown in Fig. 35 there are also differences in component latency of the maximal gradient of the pattern ERG between experimental conditions, with lower

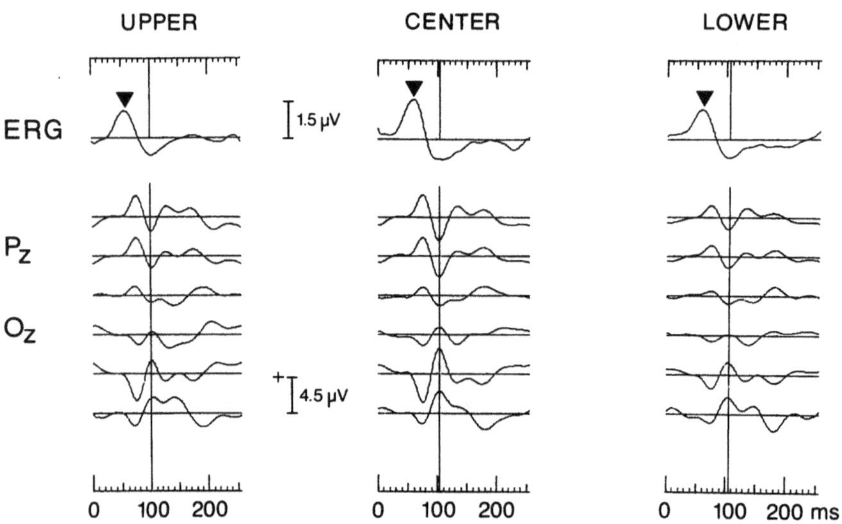

Fig. 35. Pattern ERGs and potential waveshapes evoked by a checkerboard reversal stimulus presented to the upper, the central, or the lower retina. Waveshapes in the first row show the pattern ERGs, computed as the difference between the eye electrode and the electrode at the outer canthus; the other waveshapes are the scalp data which were computed vs the average reference. *Vertical lines* at times of maximal field power. Note latency differences between upper and lower retinal stimuli in the potential gradients at the eye (component latencies between 50 and 60 ms) as well as in the scalp distribution data (component latencies around 100 ms)

retinal stimuli yielding longer latencies than upper retinal stimuli. In the subject population component latency of the pattern ERG was determined independent of the reference electrode by the occurrence of the maximal potential gradient. This resulted in a mean latency of 55.3 ms for upper retinal stimuli while the responses of the lower retinal areas averaged 59.3 ms in the subject population. The differences were statistically significant ($t = 4.69$, $p < 0.0005$). The mean component latency evoked by central stimuli was 57.5 ms, and it was different from either upper or lower component latencies only at the 10% significance level. The VEP data which had been obtained simultaneously displayed also differences in component latencies as illustrated in Fig. 35. The mean latency difference between components evoked by upper and lower retinal stimulation in the 11 subjects studied averaged 2.0 ms. This comparatively small mean difference might have been caused by the rather large interelectrode distances used and the accordingly crude sampling of the evoked scalp field which contributed to the interindividual variability. However, analysis of the individual data showed that the latency difference was consistent in the subject population, and it achieved statistical significance ($t = 2.19$, $p < 0.005$).

In summary, the present data show that localized visual stimuli can be used to investigate different parts of the retina by means of ERG recordings. Differences already observed in the interaction between the retinal pigment epithelium and the photoreceptors (Sect. 4.1) also show up in the electrical responses associated with retinal ganglion cell activity: Stimulation of upper retinal areas yields significantly shorter component latencies of the pattern ERG than stimulation of the corresponding lower retinal areas. While the significant differences observed in the pattern ERG were restricted to the time domain, as reflected by significantly different component latencies, in the EOG recordings the differences were most pronounced in the amplitudes of the light-induced standing potential of the eye. This fact might well be accounted for by the rather crude time resolution of the determination of the light peak of the EOG, which is in the order of minutes, whereas the pattern ERG component latencies are less than 100 ms.

5 Functional Differences

The aim in this section is to investigate whether the electrophysiologically determined differences between the upper and the lower retina are paralleled by perceptual differences. Most of the electrophysiological data were discussed in the previous sections mainly in formal terms of amplitudes, latencies, or scalp topography, which do not relate directly to functional aspects of visual perception. Prompted by the evoked potential results, a series of studies was made investigating visual functions such as temporal sensitivity (Sect. 5.1), contrast sensitivity and visual acuity (Sect. 5.2), and discrimination of meaningful alphanumeric stimuli (Sect. 5.3). The results will illustrate consequences on behavioral performance which arise from the physiological differences between the upper and the lower retina, and they will show that a consistent superiority of the upper retinal system can also be found on a perceptual level. This illustrates that the interpretation of the electrophysiological results described in the previous sections is not restricted to a small and highly specialized field of research, but is paralleled by behavioral concomitants.

5.1 Temporal Sensitivity[4]

The temporal resolution capacity of the visual system depends both on physical stimulus variables such as flash duration, intensity, size, and spectral composition of the test light, and on physiological and subject conditions such as state of adaptation or retinal location (for reviews see Landis 1954; Brown 1965; Shickman 1981). Differences between rod and cone functions are reflected in different critical flicker fusion frequency (CFF) values obtained in the center and the periphery of the retina. For more practical clinical applications, the visual field was mapped by Phillips (1933), Hylkema (1942), and Kleberger (1955) using the CFF. These authors described different sensitivities of various parts of the visual field.

The sensitivity to flickering light is not only determined by retinal mechanisms (Dodt 1951; Dodt and Enroth 1954; Grüsser and Kapp 1958; Grüsser and Rabelo 1958); central parts of the visual system also contribute to the CFF, as has been demonstrated by binocular stimulation experiments in humans (Simonson 1958; Lipkin 1962; Zubek and Bross 1972, 1973). This interpretation was supported by Granit (1955), who found that central electrical stimulation modulates the ganglion cell activity of the retina. In addition, Dodt (1964) showed in the rabbit that tetanic stimulation of the optic tract has a direct effect on the retinal CFF. Van de Grind et al. (1973) give a more detailed review of binocular CFF and of cortical neural responses indicating binocular interactions.

Determination of CFF thus yields a global measure of the temporal sensitivity of the visual system, and it is used in this section to investigate functional differences between upper and lower, and between nasal and temporal perifoveal retinal areas. In contrast to Hylkema (1942), who investigated the retinal periphery in order to determine the visual field, in the present experiments the perifoveal regions were examined. In a second experiment, the time thresholds for the more demanding task of double flash discrimination (DFD) were determined.

Fifteen adults between 19 and 32 years of age with normal vision participated in both experiments. Stimuli were presented by the lamp of a GRASS PS2 photostimulator (white light between 450 and 650 nm, exposure 0.16 lx/s, fused flashes 2.6 candela/m^2) which was mounted in the center of a vertical white wooden board with a hole of 2.5° diameter. Small red LEDs (9.8′ diameter) located at 5° and 10° along both directions of the horizontal and vertical meridians served as fixation points. Time thresholds were measured by the method of adjustment at these eccentricities and also with central fixation. This yielded nine measured values for each subject for both CFF and DFD.

After 15 min dark adaptation, the subject fixated monocularly, with his dominant eye, either the center of the display or one of the peripheral fixation points from a distance of 174 cm. In the CFF experiment, the subject controlled the stimulus frequency, and was asked to set the frequency so that he could not quite discriminate single flashes. In the DFD experiment double flashes occurred once per second, and the subject's task was to set the interflash interval so that double

[4] The results of this section have been published elsewhere (Skrandies 1985a)

flashes could no longer be perceived. The interflash interval was monitored on an oscilloscope screen with a time resolution between 1 and 4 ms. The CFF values were expressed as flashes per second, while DFD thresholds were measured in milliseconds. After several practice trials, four to six measures were obtained at each of the nine retinal stimulus locations in a quasi-randomized order. The repeatability of the settings was good, and mean CFF or DFD values of each location were used for statistical analysis. Corresponding locations in opposite hemiretinal areas were compared by paired t-tests.

5.1.1 Results on Critical Flicker Fusion Frequency

The data of each subject were expressed as percentages of the respective central value. Figure 36 shows the man data of 15 subjects measured in the central 10° of the retina as isopercentage lines which connect areas of equal temporal sensitivity. The zero isopercentage line runs from the lower nasal part through the center to the upper temporal part, and temporal and lower retinal areas are less sensitive than the nasal and upper areas. This plot rests on interpolated values, and statistical comparisons were performed on the differences between the nasal and temporal and between the upper and the lower hemiretinal CFF values at the retinal locations actually measured. The grand mean CFF value (over all conditions and subjects) was 19.35 flashes per second. In paired t-tests on the individual data, the differences between nasal and temporal CFF were significant only at 5° eccentricity ($t = 1.87$, $p < 0.05$), while the differences between upper and lower areas were significant at both eccentricities. The mean results (CFF values in Hz) and significance levels are summarized in Table 4. In the temporal retina the mean CFF was 92.2% of the nasal value at 5° while at 10° it reached

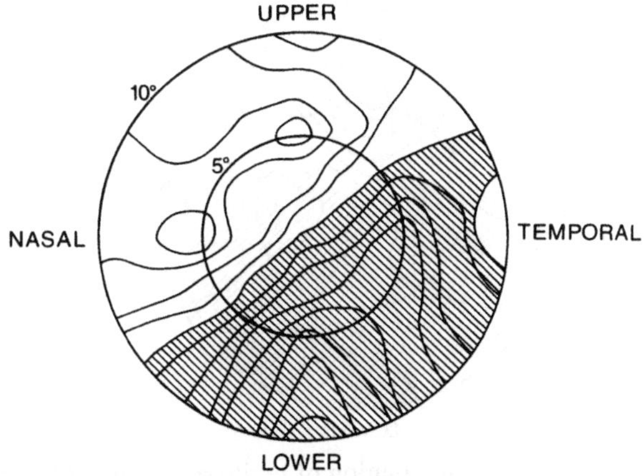

Fig. 36. Retinal distribution of perifoveal mean CFF values (%) of 15 subjects. The central value is set to 100%, isopercentage lines connect points of equal percentage values. Lines are in steps of 1%. CFF in the *shaded areas* is smaller, in the *white areas* larger than the central CFF. The *inner circle* represents 5°, the *outer circle* 10° retinal eccentricity. (From Skrandies 1985a: by permission of Gordon and Breach Science Publishers, London)

Table 4. Mean CFF values (Hz) of 15 subjects

Eccentricity					
5°			10°		
Upper		Lower	Upper		Lower
20.45		18.63	19.50		18.03
	$t = 3.40$			$t = 3.54$	
	[$p < 0.0025$]			[$p < 0.0025$]	
Nasal		Temporal	Nasal		Temporal
20.44		19.02	20.06		19.67
$t = 1.87$				n.s.	
	[$p < 0.05$]				

Paired t-tests were used for statistical analysis

97.8% of the nasal values. The mean differences between the upper and lower areas were close to 10%: at 5° the values for the lower retinal averaged 90.2% of the upper retinal CFF, and at 10° the lower retinal CFF values were 91.8% of the upper retinal CFF values.

5.1.2 Results on Double Flash Discrimination

The results obtained for DFD thresholds were similar to the CFF data. Figure 37 shows the sensitivity distribution of the 15 subjects across the perifoveal retinal

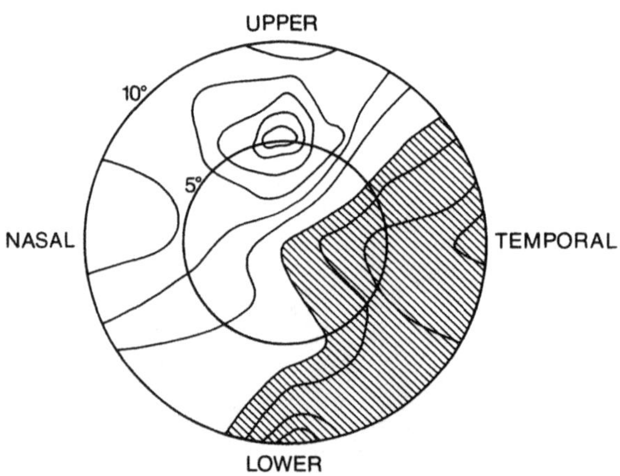

Fig. 37. Retinal distribution of perifoveal mean DFD threshold values (%) of 15 subjects. The central value is set to 100%, isopercentage lines connect points of equal percentage values. Lines are in steps of 2%. Threshold values in the *shaded areas* are smaller, in the *white areas* larger than those of the central area. The *inner circle* represents 5°, the *outer circle* 10° retinal eccentricity. (After from Skrandies 1985a: by permission of Gordon and Breach Science Publishers, London)

Table 5. Mean DFD values (ms) of 15 subjects

Eccentricity					
5°			10°		
Upper		Lower	Upper		Lower
82.75		92.80	91.69		100.91
	$t=2.83$ [$p<0.01$]			$t=2.88$ [$p<0.01$]	
Nasal		Temporal	Nasal		Temporal
88.96		97.32	87.77		100.43
	$t=2.66$ [$p<0.01$]			$t=3.04$ [$p<0.005$]	

Paired t-tests were used for statistical analysis

areas. As in the CFF data, the nasal and upper retinal areas have better time resolutions than the temporal and lower retinal areas (Fig. 37). In these data the zero isopercentage line is somewhat different from that in the CFF plot (see Fig. 36), but all differences between corresponding upper and lower locations and between corresponding nasal and temporal locations were statistically significant. The grand mean threshold (over all subjects and over all retinal locations) was 92.88 ms, and the mean DFD values (indicating the interflash intervals in milliseconds) and significance levels obtained with paired t-tests are given in Table 5. The mean differences between the retinal areas examined were in the same order as those of the CFF thresholds. Lower retinal stimulation yielded thresholds of 112% of the upper retinal values at 5° and 110% at 10°, and the temporal areas showed higher values than the nasal retina: At 5° the mean values were 109%, and at 10° 114% of the nasal values.

5.1.3 Relation Between Critical Flicker Fusion Frequency and Relative Luminance

An additional experiment was performed in order to relate the regional differences in temporal sensitivity to relative luminance thresholds. In two subjects CFF measurements were obtained using test lights at eight different luminance levels. The relation between CFF and log filter density is shown in Fig. 38. For both subjects the CFF decreased with increasing filter density, and this relationship showed a high correlation (identical value for both subjects: $r = -0.97$). Linear regression lines were determined for each subject individually, and although the subjects have different levels of temporal sensitivity, the slopes of the regression lines in Fig. 38 are identical. For both subjects a difference in CFF of 10 flashes per second corresponds to 1.36 log filter units density.
The basic results of this section are in general agreement with the data of Phillips (1933), Hylkema (1942), and Kleberger (1955); upper and nasal retinal areas show

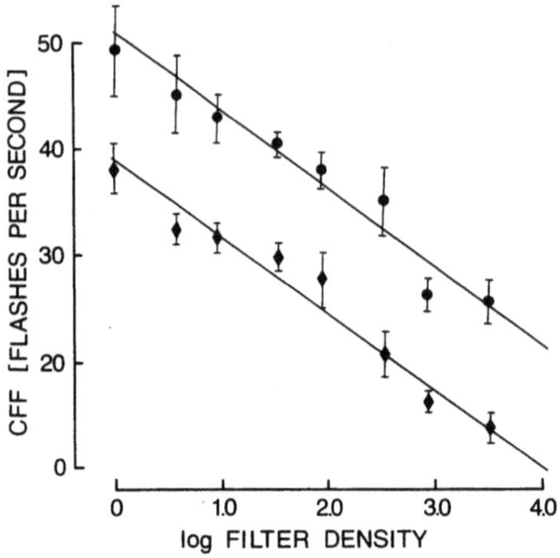

Fig. 38. CFF as a function of log filter density of two different subjects. (Filled circles or diamonds). Means and standard deviations of ten measurements are illustrated. Light intensity decreases from left to right. The lines are linear regressions computed from individual data. The correlation coefficients and slopes are identical for both subjects ($r = -0.97$, $a = 0.136$); only the absolute CFF levels are different. (Data from Skrandies 1985a)

higher resolutions than lower and temporal retinal areas (see Sect. 6.2.2.2 for a more complete discussion). However, in the earlier studies the vertical meridian was investigated much further out in the periphery, while in the present experiments significant differences in the temporal resolution capacity were found already in perifoveal retinal areas.

The data presented show a consistent superiority of the upper and nasal over the lower and temporal perifoveal retinal areas. The differences in temporal resolution were found both in the CFF and in the DFD measurements. The discrimination of double flashes yields higher time thresholds than the CFF (Basler 1911; Lindsley and Lansing 1956) because the state of adaptation is different in these two experimental conditions. After prolonged stimulation the subject becomes light-adapted and CFF increases (Granit and Hammond 1931). This was found in the present experiments: for CFF, a mean interflash interval of 51.67 ms could be resolved, while double flashes could only be discriminated at an interflash interval of 92.88 ms. In addition, all our subjects agreed that the DFD experiment was more difficult to perform than the CFF experiment. Despite this difference in the level of achievement, for the DFD task the relative size of the significant functional differences was of an order similar to the CFF. In both experiments the time resolution values of the lower retinal areas were about 10% less than those of the upper retina, and the nasal-temporal differences were of a similar magnitude.

The measured values were also expressed as relative luminance thresholds. As expected from the literature (Brown 1965; Shickman 1981), the CFF dropped with decreasing luminance of the test field, and this finding was consistent over subjects. The different CFF levels of the subjects in Fig. 38 reflect individual differences in threshold criterion. Most important, the relationship between CFF and log filter density showed a high negative correlation, and the slope of the regression lines was identical for both subjects. For the population data, the mean CFF differences between the upper and lower retinal areas correspond to 0.244 log filter density units, which means that the lower retinal areas need about 1.75 times more light in order to achieve the same temporal resolution as the upper

retinal areas. The nasal-temporal difference averaged 0.123 log filter density units, corresponding to a light intensity factor of 1.33.
These small differences are very consistent in the subject population, and the magnitude of the effect of the order of 10% still permits an uniform functioning within the perifoveal visual field. In summary, statistically significant differences exist for both the CFF and the DFD task, showing a higher temporal sensitivity of the nasal and upper than of the temporal and lower retinal areas.

5.2 Visual Acuity and Contrast Sensitivity [5]

There are reports suggesting regional differences in visual acuity between upper and lower retinal areas (Landolt and Hummelsheim 1904; Low 1943). Using Landolt rings of high contrast, Millodot and Lamont (1974) determined the visual acuity along the vertical meridian in three subjects. This section will report acuity differences between upper and lower perifoveal areas using a more complete assessment of visual resolution.
The measurement of human contrast sensitivity introduced by Schade (1956) and applied to the visibility of complex visual grating patterns by Campbell and Robson (1968) allows a more detailed examination of spatial visual resolution than conventional measures of visual acuity, which emphasize primarily high spatial frequency resolution at high contrast levels. Clinical studies on amblyopic patients (Gstalder and Green 1971) as well as studies on patients with multiple sclerosis (Regan et al. 1977) or with cortical lesions (Bodis-Wollner and Diamond 1976) demonstrated that normal visual acuity does not rule out perceptual disturbances. Brettel et al. (1982) and Lawden (1982) found that in amblyopic patients, not only contrast sensitivity but also sensitivity to spatial phase may be disrupted. In addition, Skrandies and Gottlob (1986) showed that patients suffering from Parkinson's disease also had normal conventional visual acuities but significantly reduced contrast sensitivity. The measurement of contrast sensitivity can thus be used to study functional characteristics of various retinal areas related to visual resolution over the whole range of spatial frequencies from very coarse to very fine visual targets which are believed to be analyzed by the visual system in independent spatial frequency channels (for review see Braddick et al. 1978).
In this section the contrast sensitivity thresholds of upper and lower perifoveal hemiretinal regions will be compared over a range of different spatial frequencies. In addition, we will show that these data also suggest differences in visual acuity. Twenty subjects between 19 and 32 years of age with normal or corrected-to-normal vision were presented static sine-wave-modulated vertical grating patterns of a mean luminance of 1.21 candela/m^2 on the screen of a R5030 Tektronix oscilloscope (P31 phosphor; field size 5.99°×7.40° arc). Gratings of seven different spatial frequencies were used: 0.45, 0.62, 1.24, 1.96, 4.58, 9.13, and 13.74 cycles per degree (c/deg). From a distance of 100 cm, the subject fixated monocularly, with his dominant eye, either the center of the display or one of the peripheral fixation points 5° above or below the edge of the display. An ascending method of limits was used. After several practice trials, five measurements were

[5] Results of this section have been published elsewhere (Skrandies 1985b)

obtained at each of the seven spatial frequencies. The repeatability of the settings was good as quantified by the very low standard deviations.

For each subject and each retinal location the contrast sensitivity function was determined, and the lowest value of each curve defined maximal sensitivity. An analysis of variance established a significant effect of retinal stimulus location, and the shapes of the contrast sensitivity functions were also compared statistically using paired t-tests. The significance levels of the t-values were corrected for multiple comparisons according to Holm's (1979) procedure, and all results presented were significant at an alpha level of $p<0.05$ unless stated otherwise. The values obtained at the three highest spatial frequencies were used to compute linear regressions between contrast threshold and spatial frequency for each retinal location separately. By extrapolating the regression lines to the high-frequency cutoff point (i.e., to 100% contrast) conventional visual acuity was estimated.

The grand mean contrast sensitivity functions computed from the data of 20 subjects obtained at the three different retinal locations are shown in Fig. 39A. The overall contrast levels are different, with central stimuli showing the lowest threshold values (mean over all spatial frequencies and subjects: 1.46% contrast), followed by upper retinal stimuli (mean over all spatial frequencies and subjects: 1.93% contrast). The lower hemiretina shows the highest contrast threshold values at all spatial frequencies (mean over all spatial frequencies and subjects: 2.42% contrast). The shapes of the contrast sensitivity curves were significantly different, as revealed by an analysis of variance (paired t-tests comparing retinal locations using all the data irrespective of spatial frequency also showed significant differences between the three retinal locations).

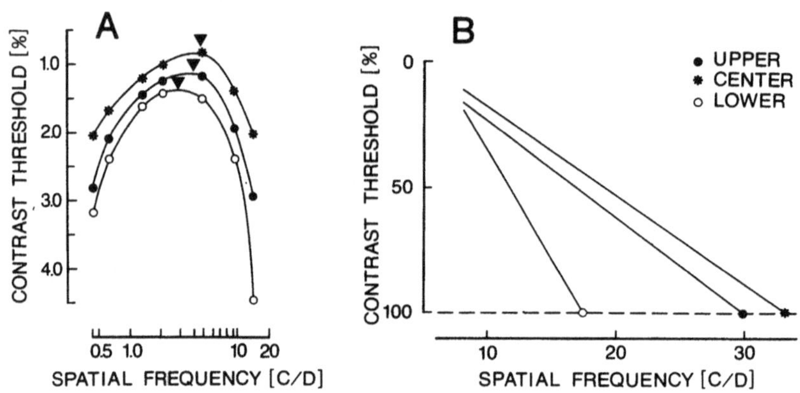

Fig. 39. A Mean contrast sensitivity curves of 20 subjects for central and for upper and lower hemiretinal perifoveal areas. *Arrowheads* indicate the mean spatial frequencies which yielded the highest individual contrast sensitivities. The differences in overall contrast level, as well as in the spatial frequencies to which subjects were most sensitive, were significantly different. (After Skrandies 1985b) **B** Linear regressions between contrast threshold and spatial frequency for upper, central, and lower retinal stimuli. Extrapolation of the regression lines to the high-frequency cutoff point (i.e., to 100% contrast) gives an estimate of conventional visual acuity. Cutoff frequencies at 100% contrast were 33.16 c/deg for central, 29.98 c/deg for upper, and 17.50 c/deg for lower retinal stimuli (see text)

The lowest contrast thresholds occur at different spatial frequencies for the different retinal locations, indicating sensitivity differences (Fig. 39A). The spatial frequencies at which each subject was most sensitive were also compared between retinal locations, and all paired t-tests on spatial frequency differences between central (mean 4.31 c/deg), upper (mean 3.85 c/deg), or lower (mean 2.63 c/deg) retinal stimulus locations were significant, as summarized in Fig. 40.

In addition, comparisons of the contrast values at each of the seven measured spatial frequencies were made: All statistical comparisons were highly significant, indicating that there are regional retinal differences not only in visual acuity, as suggested by the different contrast levels at the highest spatial frequency, but that the global differences are not restricted to individual spatial frequency channels. For the three highest spatial frequencies the correlations between contrast threshold and spatial frequency were highly significant ($p<0.005$) at all retinal locations investigated, ranging from 0.6 to 0.7. The linear regressions between contrast threshold and spatial frequency are illustrated in Fig. 39B. Central and upper stimuli yielded similar slopes of the lines, and comparable cutoff frequencies at 100% contrast (33.16 c/deg for central, 29.98 c/deg for upper retinal stimuli). The cutoff point determined for lower retinal stimuli was 17.50 c/deg. When expressing these cutoff frequencies as visual acuity, the grand mean data of Fig. 39B yield acuity values of 1.1 for central, 0.99 for upper, and 0.58 for lower retinal stimuli. Foveal stimuli consistently yielded the highest contrast sensitivities and the highest cutoff spatial frequencies (i. e., visual acuites), as well as a shift of the maximal sensitivity toward higher spatial frequencies than perifoveal stimuli. This reflects the known anatomical and functional superiority of the fovea with its dense packing of cones and its low amount of convergence onto retinal ganglion cells, enabling high spatial resolution.

The findings on upper/lower hemiretinal differences are in agreement with those of Landolt and Hummelsheim (1904), Low (1943), and Millodot and Lamont (1974), who determined the conventional visual acuity at high contrast levels. Millodot and Lamont (1974) described differences in visual acuity only beyond 5° eccentricity, and Low's (1943) data were obtained at 30° eccentricity. Wertheim (1894) had also found higher visual acuities for the upper hemiretina only beyond 20° eccentricity. In contrast to these authors, the present data show that differences between the upper and the lower hemiretinal areas exist in the perifovea at as little as 5° eccentricity (see Sect. 6.2.2.3 for a more complete discussion of the earlier literature).

The contrast sensitivity functions of the perifoveal retinal regions presented in this chapter show a significant superiority of the upper over the lower retinal areas.

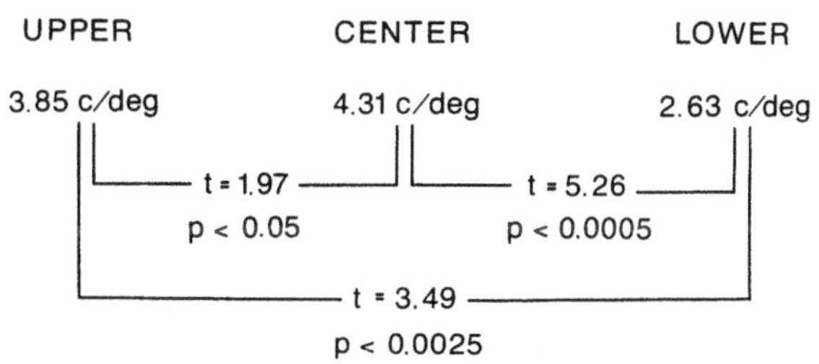

Fig. 40. Mean maximal contrast sensitivities obtained from 20 human subjects. Statistically significant differences between central, upper, and lower retinal perifoveal area were established by paired t-tests

These differences in contrast sensitivity were not restricted to high spatial frequencies or estimated conventional visual acuities, but were observed at all spatial frequencies. Differences in maximal sensitivity, as well as the different shapes of the contrast sensitivity functions, suggest a global functional superiority of the upper over the lower perifoveal hemiretinal areas. This functional difference is not restricted to certain individual spatial frequency channels.

5.3 Discrimination of Meaningful Stimuli

Pattern discrimination of meaningful stimuli can be considered a realistic visual task in everyday life. In this section consequences in behavioral performance which arise from the physiological differences between the upper and the lower retina are illustrated. In a simple pattern discrimination task, digits were presented at an eccentricity of 5° either above or below the fovea, and the subject had to report which digit he had seen. In contrast to experiments using checkerboard or grating stimuli, the discrimination of numbers presented on a computer terminal is not an artificial, laboratory-bound experimental task for the subject. Fifteen young adults (19–33 years of age) with normal vision participated in the experiments. One of ten different digits (0–9) was flashed in a randomized order on the terminal of an Apple IIe computer, 5° above or below a central fixation point consisting of a light cursor. The size of the test target subtended a 30′ arc. Both sequences of stimulus location and digits were randomized. The subjects responded by entering the appropriate number on the keyboard. After a programmed delay of 1 s, the next stimulus appeared. Thus each subject was free to work at his own pace. Each subject performed 600 trials. Accuracy in performance was stressed, and a forced-choice task was employed. For analysis, percentages of correct responses were computed and the discrimination performance of upper and lower retinal areas was compared using Wilcoxon tests. Luminance and contrast of the display were adjusted for each subject individually to yield a mean above 50% correct responses.

The performance of all subjects was significantly above the chance level of 10%. The mean of correct responses computed over all subjects and all retinal locations was 74.39% correct responses. There were no obvious differences in discrimination performance between individual digits. Because a forced-choice task was employed, confusions mainly arose between similarly shaped digits (e. g., 1,4,7 or 6,8,9,0 were incorrectly reported).

The individual data of the 15 subjects and the group means are shown in Fig. 41. In 14 of the 15 subjects stimuli presented to the upper retina yielded better discrimination performances than those presented to the correponding lower retinal area. With upper retinal stimuli 78.63% correct discriminations were obtained, whereas only 70.14% of the lower retinal stimuli were correctly discriminated. This difference was statistically significant in the subject population (Wilcoxon $T = 14$, $n = 15$, $p < 0.005$).

In this experiment, fixation was deliberately not very strictly controlled. Nevertheless, systematic significant differences in discrimination performance were found. In order to achieve a high level of accuracy, the subjects' best strategy was to fixate at the center of the screen, because the location of the expected stimulus

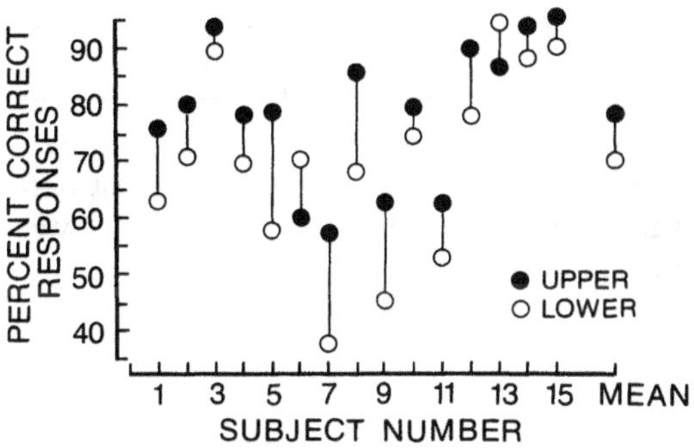

Fig. 41. Percentages of correct responses in a pattern discrimination task. Individual data of 15 subjects as well as the means are shown. Stimuli presented to the upper retina yielded significantly better discrimination performances than those presented to the corresponding lower retinal areas (78.63% vs 70.14% correct discriminations)

could not be predicted. The subjects were naive as to the exact purpose of the experiment and had been encouraged to work as accurately as possible at their own pace. As expected from the data on temporal resolution and visual acuity presented in Sects. 5.1 and 5.2, stimuli occurring below the fixation point could be discriminated significantly better than those above the fixation point. In this task, too, the stimuli were presented only at 5° eccentricity, and the upper retinal superiority was significant in the subject population. In addition, we note that the differences between upper and lower retinal performance were again in the order of 10%.

Related differences in detection performance between visual targets presented above or below the fixation point were found in a study on attention and visual search using a much more complex "contingent-display technique". Those results were interpreted as an interaction between attention and sensitivity across the visual field (Prinz 1984; see also Sect. 6.2.2.4). Our data illustrate how, in a simple experimental setup, functional differences between upper and lower retinal areas may be demonstrated. The results were obtained in a common visual task and in a realistic instrumental suround, and they might be of interest in the investigation of man-machine interactions in everyday working conditions.

6 Evidence from the Literature

The aim of this section is to provide further evidence of the upper/lower distinction within the visual system by giving background information on what is known from the scientific literature on this topic. It cannot be an exhaustive review, mainly because the differences between the upper and lower visual systems have rarely been investigated directly. Most of the data presented is found in papers concerned with other research aims. It is important to note that this review will not only illustrate that the upper retina is superior over the lower retina (whatever this means in functional, perceptual terms), but also give interesting examples and experimental data which suggest that to a certain degree the upper and lower parts of the retina and the visual system as a whole may be considered independent systems. On the other hand, I do not mean to challenge the integrative action of the visual system.

In order to organize the review, a somewhat arbitrary classification of the literature had to be made, and structural (i. e., anatomical) and functional local differences within the visual system will be discussed separately. The first part (Sect. 6.1) is organized according to the anatomy of the visual system, describing regional differences at the retina, at midbrain structures, and at the cortex of animals and humans. Functional differences will be described in the second part (Sect. 6.2), which reviews findings in animals and humans, while in the last part (Sect. 6.3) clinical observations will further illustrate that the upper and lower visual systems may behave as independent units.

6.1 Anatomical Findings

6.1.1 Animal Data

6.1.1.1 Retina, Optic Nerve, and Optic Tract

There is some anatomical evidence of differences between upper and lower areas in the eye. Dogs and cats have a tapetum beneath a portion of their retina, and the pigment epithelium cells overlying the tapetum are not pigmented. The tapetum presumably improves scotopic vision by reflecting the incoming light, and it has been reported that the tapetum lucidum is restricted to the upper retinal half of the cat eye (Bishop et al. 1962; see also Fig. 1 of Niemeyer 1975).

In some fish and amphibians one finds very elaborate adaptations of the visual system to the environment of the animal. The four-eyed fish (*Anableps anableps*) is capable of both aquatic and aerial vision. Its eyes each have two pupils, one oriented toward the upper and one toward the lower visual field, and contain two major visual axes (see Fig. 4.5 of Ali and Klyne 1985). In addition, the retina is divided into a dorsal (upper) and a ventral (lower) part which contain different photopigments. In the ventral part of the retina, which subserves aerial vision, rhodopsin is predominant, while porphyropsin is found mainly in the dorsal, aquatic retina. Rhodopsin and porphyropsin have different spectral absorption spectra (502 nm and 522 nm, respectively), and the upper retina has an increased relative sensitivity toward longer wavelengths. This may be important for aquatic vision, since turbid water scatters shorter wavelengths more strongly than longer wavelengths (Luria and Kinney 1970).

A similar difference in the distribution of retinal photopigments is found in the bullfrog. In the larval stage it inhabits an aquatic environment, and its retina contains mainly porphyropsin. In the adult bullfrog the dorsal retina contains mainly porphyropsin, while rhodopsin predominates in the ventral retina (see Figs. 4 and 5 of Reuter et al. 1971). This might well relate to the fact that the dorsal retina more often looks down into the water whereas the ventral retina more often subserves aerial vision.

The regional distribution of the photoreceptors of the eye has been analyzed by Landau and Dawson (1970). They published some material on the retina of the sea lion, where more photoreceptors were encountered in the upper than in the lower half (see Fig. 3 of Landau and Dawson 1970).

The visual field of most birds is largely monocular, only a very small area being binocular. The binocular field is projected onto a region in the upper temporal quadrant of the retina, the so-called red area. In this region of the retina the inner limbs of the cones contain red oil globules, as found in ducks, chickens, and pigeons (King-Smith 1969). On the basis of this different distribution of red oil droplets, the wavelength discrimination capability in the lower visual field is expected to be shifted toward longer wavelengths compared to the discrimination capability in the upper visual field (see Sect. 6.2.1). The red area has a large magnification factor in the tectum which constitutes a major central visual area in the bird. In addition, in this area there is a high density of retinal ganglion cells, suggesting a greater importance of the upper temporal quadrant of the bird retina.

The cell density of the ganglion cell layer of the chick retina was studied by Ehrlich (1981). The highest value, 30000 cells/mm^2, was observed in the center, and cell density dropped sharply in all directions from the central area except for a strip in the superior temporal retina (see Fig. 9 of Ehrlich 1981). Thus, two areas with enhanced visual resolution corresponding to regions of high ganglion cell density exist in the lateral and inferior frontal visual field.

In a quantitative analysis of cat retinal ganglion cell topography, Hughes (1975) showed that ganglion cell density is higher in upper than in corresponding lower retinal areas (see Figs. 7 and 9 of Hughes 1975). In addition, differences in retinal surface area exist along both the horizontal and vertical meridians, with the upper and nasal retina occupying a larger percentage of the total retinal area than the lower and temporal areas. The upper retina represents 60% and the lower retina 40% of the total retinal area, while the nasal retina represents 72% and the temporal retina only 28% of the total retinal area (see Table 5 of Hughes 1975). According to these data the lower retina of the cat contains 98700 ganglion cells and the upper 118500, corresponding to 45.4% and 54.6% of the total ganglion cell population respectively.

Vaney and Hughes (1976) published data of a cell count in the rabbit retina and reported differences in ganglion cell density in the peripheral retina. For areas above the visual streak 550 cells/mm^2 were reported, while areas below the visual streak contained only 350 cells/mm^2.

Hughes and Wässle (1976) counted the nerve fibers and determined their diameter spectrum in the cat optic nerve. In a section about 2–3 mm behind the globe they found temporally a region with a high mean density of about 150000 fibers/mm^2. This region was assumed to correspond to the area centralis of the retina, and it extended further into the upper part of the optic nerve. A more homogeneous distribution was observed some 7 mm behind the globe, where the densities in the upper half were higher than those in the lower half of the optic nerve. Although the retinotopic order in the optic nerve is rather crude, these data agree well with the findings on different ganglion cell densities in the upper and lower retina presented above.

The topography of amacrine cells in the cat retina also shows differences in the distribution along the vertical meridian. Upper retinal areas display a higher cell density than corresponding lower retinal areas, and the density gradients appear to decrease continually from the center toward the periphery in the lower retina while in the upper retina a plateau of equal cell density is observed (see Fig. 2 of Schmidt et al. 1985; Vaney 1985). On the other hand, Oyster et al. (1985) reported that the distribution of dopaminergic displaced amacrine cells as well as that of dopaminergic interplexiform cells showed highest cell densities in the inferior part of the cat retina (see Fig. 2 of Oyster et al. 1985).

The topography of horizontal cells of the cat retina was examined by Wässle et al. (1978). Cell density was reported to be highest in the center and to decrease toward the periphery. Inspection of the isodensity lines reveals an asymmetrical distribution of horizontal cells in the upper and lower retina, with the upper retina showing higher cell densities (see Fig. 9 of Wässle et al. 1978). The dendritic fields of the horizontal cells were found to increase from the center toward the periphery of the retina, and the dendritic field diameters appear to be slightly smaller in the upper than the lower retina, suggesting a smaller amount of convergence of photoreceptors onto retinal ganglion cells (see Fig. 10 of Wässle et al. 1978).

There is a rough retinotopic arrangement of the nerve fibers forming the cat optic tract with a representation of the lower visual field in its dorsolateral part (Torrealba et al. 1982). In an electrophysiological study Aebersold et al. (1981) investigated the representation of the visual field in the optic chiasm and the optic tract of the cat. They recorded the electrical activity of 749 fibers from 19 adult cats. Fibers originating in the upper hemiretina were much more frequently encountered than fibers from the lower retina. This ratio averaged 63%: 37% and was similar in the optic chiasm and the optic tract. From these data the authors concluded that the upper hemiretina is more strongly represented in the visual system and that it presumably possesses a higher retinal ganglion cell density than the lower hemiretina.

6.1.1.2 Midbrain Structures

In various species, differences between upper and lower areas of the visual field have been demonstrated even at the level of the midbrain.

The pigeon retina receives centrifugal nerve fibers terminating on amacrine and displaced ganglion cells in the retina. These fibers originate in the isthmo-optic nucleus, where Holden and Powell (1972) described a greater representation of the lower visual field and particularly of the red area of the retina. The visual field is mapped retinotopically on the optic tectum, with larger areas occupied by the fovea and the red area (Clarke and Whitteridge 1976). The wulst of the pigeon, reminiscent of the mammalian visual cortex, is retinotopically organized and has a columnar arrangement. Most visually active units in various layers of the wulst were found to have their receptive field centers in areas corresponding to the fovea and the red area of the retina (Gusel'nikow et al. 1976, cited by Emmerton 1983a). The overlapping binocular visual field of the pigeon is located in the inferior frontal field and corresponds to the red area (for review see Emmerton 1983b).

According to Whitteridge (1973) the horizontal meridian divides the surface of lamina A of the cat lateral geniculate nucleus (LGN) unequally. The lower visual field is projected to an area which is about twice as big as that devoted to the upper visual field. In particular, the visual field just below the horizontal meridian and up to an eccentricity of about 10°–20° is extensively represented. Such differences between the upper and lower visual fields in the LGN of the cat, with the lower visual field occupying a larger area than the upper visual field, had been reported earlier by Whitteridge and Seneviratne (1962) and by Seneviratne (1963).

In small animals like the rat, mouse and frog, some evidence has been obtained that an asymmetry exists in midbrain structures of the visual system, favoring the upper part of the visual field.

Nagata and Hayashi (1984) investigated the projection of the visual field onto the ventral lateral geniculate nucleus (LGNv) of the rat. This structure shows a retinotopic organization where the nasal visual field is represented dorsally and the temporal visual field ventrally. The upper visual field projects to rostrolateral and the lower visual field to caudomedial parts of the LGNv. Nagata and Hayashi (1984) recorded electrical activity from 194 cells of the rat LGNv, and analyzed the distribution of the corresponding receptive field centers in the visual field. They found a more extensive representation of the upper visual field (up to 40° eccentricity) than of the lower visual field (up to 20° eccentricity), but there was a higher cell density representing the lower visual field. This difference in cell density was most evident when comparing receptive field centers in an area up to 20° above the horizontal meridian with those in the corresponding area extending to 20° below the horizontal meridian (see Figs. 3 and 4 of Nagata and Hayashi 1984).

A similar asymmetry between upper and lower visual fields exists in the mouse visual system. The visual projections to the mouse superior colliculus were mapped with microelectrode recordings by Dräger and Hubel (1976) in seven animals. It was found that the upper visual field is disproportionally represented (see Fig. 2 of Dräger and Hubel 1976). An area of high magnification was reported some 30° above the horizontal meridian, indicating that a larger amount of tectal surface is devoted to a point in the upper than in the lower visual field.

The nucleus isthmi of the frog visual system shows an orderly topographical organization, as has been demonstrated using neuroanatomical techniques (Scalia 1976), and an interesting functional difference between the upper and lower visual field projections to the frog nucleus isthmi was reported by Shurong et al. (1981). They recorded visual responses of 125 neurons and confirmed the retinotopic representation of the visual field. Only a small number of binocular neurons were found, mainly located in the medullary part of the rostral half of the nucleus (Fig. 4 of Shurong et al. 1981). This part of the nucleus isthmi receives projections from lower retinal areas, and it appears conceivable that this finding is a physiological correlate of the greater importance of the frog binocular upper visual field.

6.1.1.3 Cortical Areas

An orderly retinotopic mapping was reported for most cortical visual areas in all species studied. There are several cortical areas involved in visual information processing, and among the best studied are the primary visual cortex (area 17 of Brodmann) and the circumstriate belt (areas 18 and 19). Other visual cortical areas exist and have been investigated, and the functional topography of the visual cortex has been summarized by Creutzfeldt (1983) and van Essen (1985). In this section, mainly differences in the cortical representation of the upper and lower visual fields will be reviewed. The reported systematic differences in the amount of cortex devoted to the upper or lower hemiretinal projections, as well as the differences in neuronal response properties in extrastriate cortical areas representing the upper or lower visual field, yield a rather consistent picture.

Bilge et al. (1967) mapped the visual cortex of the cat by recording evoked potentials and single unit activity. One of their main findings was that the cortical surface area devoted to the upper visual field (i. e., the lower hemiretina) was considerably smaller than that devoted to the lower visual field. This was true for both area 17 (visual I) and area 18 (visual II).

In area 17 of the cat, extensive microelectrode mapping experiments were performed by Tusa et al. (1978). They found that a larger cortical area is devoted to the lower than to the upper quadrant of the visual field (see Fig. 6 of Tusa et al. 1978).

Van Essen et al. (1984) studied the visual field representation in the striate cortex of six macaque monkeys, presenting microelectrode data from a total of 156 recording sites. From their summary of all data obtained it is clear that the upper hemiretina is distributed over a larger cortical area than the lower hemiretina (see Fig. 4 of von Essen et al. 1984). This difference was most pronounced for the visual fields within 45° above or below the horizontal meridian.

Standage and Benevento (1983) demonstrated a topographical projection onto the visual midtemporal area (MT), which is located in the superior temporal sulcus of the macaque monkey, and van Essen et al. (1981) found that upper and lower visual fields are represented asymmetrically in this area. Only a total of 26% of the receptive fields analyzed had their centers above the horizontal meridian, and there was also underrepresentation of receptive field centers close to the upper vertical meridian.

Some of the intrahemispheric projections in the macaque monkey visual cortex are different for the upper and lower visual fields. The dorsal part of the striate cortex (V1), which is connected to the upper hemiretina, projects to visual area V3, while the ventral part of V1 representing the lower retina shows no corresponding projection (van Essen et al. 1986).

In addition, there are differences in physiological response properties of neurons in extrastriate cortical areas representing the upper or lower quadrants of the visual field. In the macaque visual cortex the area V3 located in the dorsal extrastriate cortex contains a representation of the lower visual field, while the ventral posterior area (VP) is located in the ventral extrastriate cortex and represents the upper visual field. Burkhalter et al. (1986) reported that in V3 a larger percentage of neurons were direction-selective, while in VP more neurons were color-selective. This indicates that there are functional asymmetries relating to the processing of information about motion and color in extrastriate cortical areas which represent the upper or lower visual field.

A double representation of the lower visual quadrant was reported to exist in the prelunate gyrus of the rhesus monkey (V4 area). Baizer and Maguire (1983) analyzed single unit activity obtained at 358 recording sites in five hemispheres of three monkeys. Their main finding was that most of the V4 complex consisted of two visual areas in the posteromedial and the anterolateral parts. These areas were retinotopically organized, and both represented the contralateral, lower visual quandrant. The receptive fields found were very large, and Baizer and Maguire (1983) concluded that the neurons in the V4 area do not contribute to the spatial visual resolution; rather, these neurons are concerned with more "global" aspects of vision.

There are also efferent connections from the visual cortex to subcortical areas, and the corticoponto-cerebellar pathways are believed to be involved in the guidance of movement (Glickstein et al. 1972). In cats, the visual areas 17, 18 and

19 have been shown to project to rostral pontine nuclei which in turn project to the cerebellum. The corticopontine neurons in area 18 are binocularly driven, have large receptive fields, and they are directionally sensitive and respond to downward movement at high velocities. Cells in the rostral pons show similar behavior and their large receptive fields respond to downward movements (Gibson et al. 1978). This organization enables the coding of movement of the terrain and gives the animal information about its locomotion. For such functional coordination of the visual input with motor performance, the lower parts of the visual field are most important. This is reflected in the results of anatomical studies on the corticopontine connections: In the cat visual cortex it has been demonstrated that in all visual cortical areas studied, the cortex representing the lower visual hemifield has a significantly higher number of corticopontine cells than the cortex representing the upper hemifield (Bjaalie and Brodal 1983; Bjaalie 1985). The corticopontine cells of the visual cortex are presumably involved in the transmission of spatially relevant information, and Bjaalie and Brodal (1983) showed in four cats that a mean of 80.5% of all cells labeled with horseradish peroxidase were found in parts of area 17 which represented the lower visual field. In a recent study on the distribution of corticopontine cells in areas 18 and 19 of the cat, Bjaalie (1985) found a similar overrepresentation of the lower visual field in these cortical areas.

6.1.2 Human Data

6.1.2.1 Retina

Most of the data on the anatomy of the mammalian retina have been obtained from the cat, and there have been few studies on the human retina. Østerberg (1935) gave a quantitative account of the distribution of rods and cones in the retina of a 16-year-old boy. From this quantitative study Østerberg (1935) concluded that there are higher densities of both classes of photoreceptors in the upper and nasal than in the lower and temporal parts of the retina (Figs. 6 and 7 of Østerberg 1935).

A recent report on retinal punches obtained at 20° and 40° eccentricity along the vertical and horizontal meridians of the human retina confirms a higher density of photoreceptors in the nasal and upper than in the temporal and lower retinal areas (Curcio et al. 1985).

Van Buren (1963) published data on the distribution of retinal ganglion cells along the major meridians. His data obtained from six human retinae are summarized in Fig. 42, which shows the distribution of ganglion cells along the vertical meridian. Two striking differences between the upper and lower retinal areas may be seen: First, the decay of ganglion cell densities is steeper in the lower retina. From about 10° eccentricity on, the lower retina displays a smaller ganglion cell density. The overall number of ganglion cells averaged 2325058 for the upper meridian while along the lower retinal meridian 2137840 ganglion cells were counted (van Buren 1963, p. 64).

Second, there is a difference in the amount of overlapping ganglion cell layers (Fig. 42). More centrally, up to five layers are found, but toward the periphery the

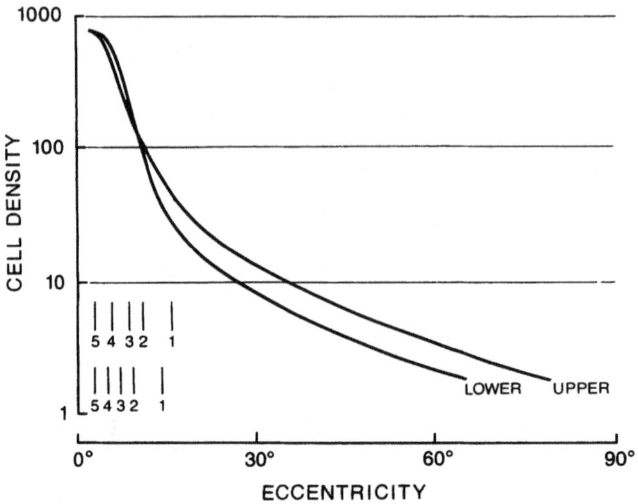

Fig. 42. Ganglion cell distribution in the human retina along the vertical meridian. Cell densities in cells per square micrometer. Numbers indicate the margins of cell layers of various thickness. (Measured from Fig. 36 of van Buren 1963)

Table 6. Margins of retinal ganglion cell layers of various thickness in the human retina giving retinal eccentricities in degrees of visual angle. (Data measured from Fig. 36 of van Buren 1963)

Number of cell layers	Upper retina	Lower retina
1	15.66°	14.50°
2	11.00°	9.66°
3	9.00°	7.50°
4	6.33°	5.00°
5	3.33°	2.66°

ber of layers is reduced to one (see also Hebel and Holländer 1983). Analysis of Fig. 42 indicates that in the upper retinal areas the overlapping of ganglion cell layers extends much further out toward the periphery than in lower retinal areas. The eccentricities of the margins between the five ganglion cell layers were measured from van Buren's (1963) original data, and the results are presented in Table 6. All layers show a larger extension toward upper retinal areas, indicating a higher ganglion cell density in the upper than in the corresponding lower hemiretinal areas.

A related set of data on the distribution of ganglion cells across the human retina has been presented by Stone and Johnston (1981). They measured the ganglion cell density in one human retina, and displayed their results in the form of equidensity maps. The ganglion cell density along the vertical meridian was measured from their distribution map (Fig. 8 of Stone and Johnston 1981), and is illustrated in Fig. 43. Similar to van Buren's (1963) data, higher cell densities are

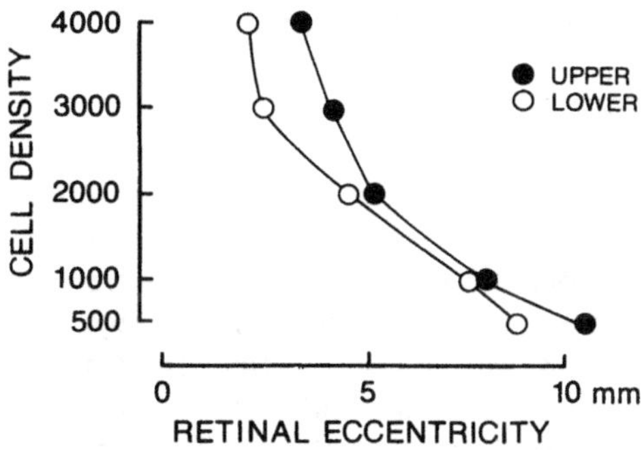

Fig. 43. Ganglion cell density along the vertical meridian of one human retina. Note consistently higher ganglion cell densities in the upper retina. (Data measured from Fig. 8 of Stone and Johnston 1981)

found in the upper retinal areas. This finding was not restricted to the areas along the vertical meridian; higher cell densities for the upper retina were also found in more nasal and temporal retinal areas (see Fig. 8 of Stone and Johnston 1981).

6.1.2.2 Cortical Areas

As in other higher mammals, the major part of the primary visual cortex of man is located on the medial aspects of the occipital cerebral hemispheres. Most of this primary visual area is folded into the calcarine fissure, and the fibers of the optic radiation representing the upper part of the visual field (i. e., the fibers originating in the lower retinal area) terminate below the calcarine fissure, while the lower visual field (i. e., the upper retina) is represented in the upper lip of the calcarine fissure.

Much of what is known about the cortical projections in man has been established in patients with localized lesions of the visual cortex (e. g., Holmes 1945; Spalding 1952; Teuber et al. 1960). The topographical mapping of the visual field onto the visual cortex is weighted by the functional importance of the retinal areas, which results in a more extensive representation of the foveal region. This has become known as cortical magnification (Whitteridge 1973).

Brindley and Lewin (1968) showed that electrical stimulation of an array of electrodes implanted over the calcarine cortex in a blind patient resulted in visual sensations whose topographies were consistent with the established maps of the visual cortex.

Newman et al. (1984) described two patients with altitudinal hemianopia caused by occipital infarctions. Computer tomograms confirmed a direct correlation of the visual field defect with the site of infarction: Defects in the superior visual field were associated with lesions in the lower bank of the calcarine fissure, while a patient with a lower visual field defect showed infarctions in the upper banks of the calcarine fissure.

Recent technological advances have made it possible to map cortical areas of the human brain also in terms of their functional activity. Phelps et al. (1981), Greenberg et al. (1981), and Schwartz et al. (1984) described the metabolic mapping of functional activity of the human brain following visual stimulation, and their data are in line with the cortical projections of the visual field described in the earlier morphological and clinical studies.

6.2 Functional Differences

6.2.1 Animal Data

Visually guided behavior was examined in cats in which one eye was rotated surgically between 9 and 15 days of age (Presson et al. 1983). Perimetry testing was performed when the cats were 6 months of age. The visual fields of the rotated eyes were abnormal, and regardless of the direction of eye rotation the cats appeared to be blind in the contralateral lower visual quadrants. This surprising finding, which was independent of the direction of rotation, could not be explained by the authors with any theory of rewiring of the visual inputs to the superior colliculus or the visual cortex.

In a study on visual neglect caused by section of the midbrain commissures in the cat, Matelli et al. (1983) demonstrated that after sections of the commissures between the superior and inferior colliculi the lesioned cats showed a pronounced inattention for stimuli presented in the upper visual field. They also lacked exploratory head movements toward the upper visual field. In contrast, a lesion of the posterior commissure resulted in a neglect of visual objects below the horizontal meridian (Matelli et al. 1983).

In a study on visually guided pecking in the pigeon using high-speed cinematography, Goodale (1983) determined where on the retina the visual target fell during the complex sequence of pecking. During periods of fixation, the target was located in the binocular visual field which corresponds to the red area of the upper temporal retina.

In behavioral experiments on pigeons, different spectral sensitivity functions were obtained by Martin and Muntz (1979) in the upper and lower visual fields: For short wavelengths below 550 nm the upper parts of the retina corresponding to the red area was found to be less sensitive than the lower parts of the retina. A similar shift of the pigeon spectral sensitivity toward longer wavelengths in the lower visual field was reported by Nuboer and Wortel (1985). This shift results in an improved discrimination capability of the animal for red and brown stimuli occurring on the ground. The physiological basis for the different spectral sensitivity of the upper and lower retina is the different regional distribution of the red oil droplets in the inner limbs of the pigeon retinal photoreceptors (see Sect. 6.1.1.1).

The pigeon visual system is also capable of detecting the plane of polarized light (Delius et al. 1976), and this enables the bird to discriminate skylight polarization. The pattern of polarization of skylight is correlated with the position of the sun, and thus gives important cues for orientation. Both behavioral and ERG data suggest that the ability to detect the plane of polarized light is associated more with the lower retina (Delius et al. 1976). These findings appear ecologically meaningful, since skylight comes from above the animal, and mainly the lower retina needs this functional advantage.

6.2.2 Human Data

6.2.2.1 Motor Reaction Time to Visual Stimuli

It has long been known that there are functional differences between retinal areas in performance of a reaction time task. A widely distributed textbook on experimental psychology mentions in its first edition in 1938 that the reaction time to visual stimuli is shorter when they are presented to upper retinal areas than when they are presented to lower retinal areas (Woodworth 1938, p. 327–328). In later editions of the same book these findings on differences between upper and lower retinal areas are omitted (Woodworth and Schlossberg 1955). Obviously, by that time the functional differences between the upper and lower retinal areas had become too well known to need further discussion.

In most studies, regional retinal differences were reported for stimuli at locations along the horizontal meridian. In has been established that reaction times are shorter to stimuli falling onto nasal than onto temporal retinal areas (Hall and von Kries 1879; Poffenberger 1912; Rains 1963).

Hall and von Kries (1879) investigated the reaction time to stimuli along the horizontal and the vertical meridian. Their data obtained at 30° and 60° eccentricity are summarized in Fig. 44. Central retinal areas yield the shortest reaction times, and differences are seen between nasal and temporal and between upper and lower retinal areas. The average reaction times were significantly shorter in upper than in lower retinal areas: 34 ms at 30° eccentricity and 46 ms at 60° eccentricity.

Differences in the same direction were reported by Payne (1965), who investigated the reaction times at 15° eccentricity in two subjects. Cocito et al. (1977) showed that the upper/lower differences depended on the spatial frequency of the stimuli used: With 1 c/deg stimuli, upper retinal areas yielded significantly shorter reaction times than lower retinal areas, whereas with 3 c/deg stimuli the authors re-

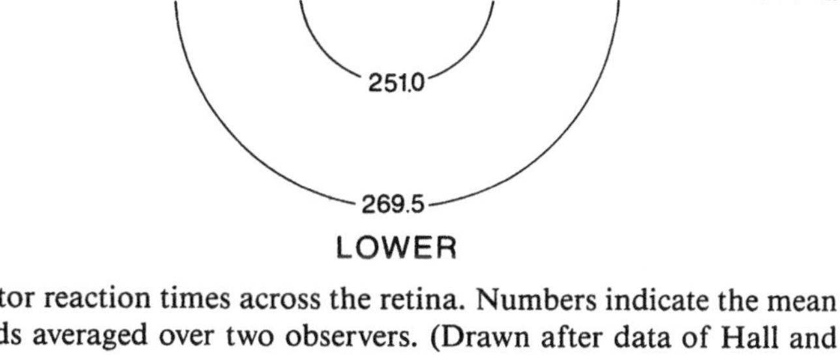

Fig. 44. Distribution of motor reaction times across the retina. Numbers indicate the mean reaction time in milliseconds averaged over two observers. (Drawn after data of Hall and von Kries 1879)

ported the same tendency but without statistical significance. Unstructured visual stimuli were used by Magrotti et al. (1980), who found significantly shorter reaction times for upper than for lower retinal stimuli at 20° eccentricity but not at 5° eccentricity.

Lichtenstein and White (1961) derived equations which permitted establishment of a relation between motor reaction time to stimuli presented in various retinal locations and the local densities of the photoreceptors measured by Østerberg (1935). Basically, these equations show that reaction time can be considered an inverse function of photoreceptor densities in the retina.

6.2.2.2 Temporal Sensitivity

As described in Sect. 5.1 our results on critical flicker fusion frequency (CFF) and double flash discrimination (DFD) are in agreement with earlier studies. The visual field plot by Hylkema (1942) illustrated that upper and nasal retina areas had higher temporal resolutions than lower and temporal retinal areas. However, Hylkema investigated the vertical meridian much further out in the periphery (at 25° eccentricity), where he obtained statistically significant differences between the upper and lower areas [$t = 2.14$, $p < 0.05$ computed from Hylkema's data (1942, p. 187)]. In Hylkema's data the lower retinal CFF reached a mean value of 96.9% of that of the upper retinal CFF. Our results show that there are significant differences in temporal resolution capacity even in perifoveal retinal areas.

A comparison of the data in the published literature reveals conflicting results concerning regional differences of CFF values. Phillips (1933) measured the CFF along the diagonal meridians from 10° to 40° in steps of 10°. His data show a tendency of the upper retina areas to have higher CFF values than the lower retinal areas, while the nasal-temporal differences appear not to be significant (Phillips 1933, Fig. 2). Kleberger's (1955) data from 30 normal subjects measured at 10°, 20°, 30° and 40° along the vertical, horizontal and diagonal meridians show no consistent differences between upper and lower retinal areas, and nasal retinal areas have an inconsistent superiority over the corresponding temporal areas. Contrary to this finding, Hylkema (1942) described at 25° eccentricity a higher CFF in the temporal retina which is thought to compensate for the nasal area of the blind spot in the opposite eye. Beyond 30° eccentricity, a nasal superiority was found (Hylkema 1942). The findings of a superiority of nasal over temporal perifoveal retinal areas, presented in Sect. 5.1, are in line with an ecological interpretation suggesting a greater functional importance of the temporal visual field, where the detection of peripherally moving and approaching targets is achieved. The nasal superiority is also reflected in the regional distribution of photoreceptors in the human retina (Østerberg 1935).

Our measurements were also expressed as relative luminance threshold values (see Fig. 38). CFF drops systematically with decreasing luminance of the test field (Brown 1965; Shickman 1981), and the mean CFF differences between the upper and lower retinal areas corresponded to 0.244 log filter density units. This means that the lower retinal areas needed a light intensity about 1.75 times higher in order to achieve the same temporal resolution as the corresponding upper retinal areas.

6.2.2.3 Visual Acuity and Contrast Sensitivity

Visual acuity appears to be directly correlated with the regional distribution of photoreceptors across the retina. Aubert and Förster (1857) investigated the performance of two subjects in a two-point discrimination task. Reanalyzing their data (Aubert and Förster 1857, Table 4) along the vertical meridian yields a significantly better performance for the upper than the lower retinal areas. In the upper retina the two-point discrimination was successful much further out in the periphery than in the lower retina. Landolt and Hummelsheim (1904) reported a replication of these experiments. These authors too found a superiority of upper over lower retinal areas (see Figs. 198 and 199 of Landolt and Hummelsheim 1904). They also mentioned the influence of the size of the target used on the determination of the visual field. For upper and nasal areas the target size had relatively little influence, whereas in the temporal and lower visual field the use of different target sizes for perimetry yielded different extensions of the visual field (Landolt and Hummelsheim 1904, p. 543).

Wertheim (1894) used grating stimuli to determine visual acuity of different retinal areas and found a higher visual acuity of the upper retina only beyond 20° eccentricity (Fig. 5 and Table on p. 182 of Wertheim 1894). In addition, he reported that the decay in visual acuity from the fovea toward the periphery was steepest over the lower retinal areas.

Landolt rings of high contrast were used by Low (1943) and by Millodot and Lamont (1974). Low (1943) investigated the visual acuity of 100 subjects and observed a better performance for upper retinal areas than for the corresponding lower retinal areas at 30° eccentricity. Millodot and Lamont (1974) mapped in three subjects the visual acuity along the vertical meridian in steps of 5° from the fovea to 40° in the periphery. They found a consistent superiority of the upper hemiretina over the lower hemiretina at locations beyond 5° eccentricity. The disagreement in these early studies concerning where exactly the differences in visual acuity between upper and lower retinal areas become detectable can be accounted for by the different stimulus parameters (e. g., visual targets, luminances, contrast) used in those studies.

Contrast sensitivity was measured in ten subjects between 61 and 71 years of age in the center and at 10° eccentricity above or below the fovea, using static and dynamic sinusoidal gratings, by Lundh et al. (1983). Similar to our results (Sect. 5.2) the fovea showed the highest contrast sensitivity, followed by the upper and then the lower hemiretina. These data, obtained further out toward the periphery than ours, are in good agreement with the findings of Wertheim (1894) and Millodot and Lamont (1974).

The contrast sensitivity functions of the perifoveal retinal regions at 5° eccentricity presented in Sect. 5.2 showed a consistent superiority of the upper over the lower retinal areas, and statistically significant differences in contrast sensitivity were observed at all spatial frequencies. Both the differences in maximal contrast sensitivity, which occurred at higher spatial frequencies for upper than for lower retinal areas, and the overall differences between the contrast sensitivity functions suggest a general functional difference between upper and lower perifoveal retinal areas which is independent of individual spatial frequency channels. Conventional visual acuity determined from these data showed large differences between the upper and lower retinal areas investigated (see Fig. 39B).

The basis for visual acuity is believed to be reflected in the mosaic of the photoreceptors in the retina, and several authors have compared the data on visual acuity with the local morphological characteristics of the retina. From these studies it was concluded that the densities of the photoreceptors or the ganglion cells in the retina show a satisfactory high correlation with the human visual acuity in various parts of the visual field (ten Doesschate 1946; Drasdo 1977; Dawson and Maida 1984). This is supported by the fact that the acuity findings presented are in agreement with the anatomical data of the human retina (see Sect. 6.1.2.1).

6.2.2.4 Perception

In a study on figural aftereffects, Freeman (1964) investigated in 12 subjects the influence of retinal stimulus location on the size of the effect. Aftereffects were obtained only when the stimulus was presented below the horizontal meridian, and as eccentricity increased, the magnitude of the aftereffect decreased. There was no statistically significant interaction between amount of eccentricity and direction on the retina (above or below the horizontal meridian), and Freeman (1964) concluded that the effect of direction of eccentricity was a general one, independent of the amount of eccentricity. This finding suggests a larger sensitivity of the human upper visual system. In a similar line, Jampolsky (1978) reported results on visual afterimages: When equal stimulus lights fall on the upper and lower retina, the perception of an afterimage on the upper retina is dominant over that of the lower retina.

The direction of the light source is important for visual perception: Changes in depth perception of convex and concave objects occur depending on the direction of light. When a picture showing indentations and bumps is inverted, indentations and bumps reverse. Our visual system is accustomed to light coming from above, and the shading gives information on convexity or concavity. This is the basis for the convex-concave illusion (cf., Teuber, 1974).

An interesting anisotropy between the upper and lower retinal systems was reported for stereoscopic depth perception. Breitmeyer et al. (1975) investigated the detection duration thresholds for dynamic random dot stereograms. They found that stimuli with uncrossed disparities (i. e., when the target was located behind the fixation plane) were detected at consistently shorter durations when presented to the lower hemiretina than when presented to the upper hemiretina. For crossed disparity stimuli (i. e., for targets closer to the subject than the fixation plane) the preference was reversed, with a superiority of the upper hemiretina. Similarly, the spatial resolution for uncrossed disparities was found to be better in the lower retina, and that for crossed disparities was better in the upper retina. Thus, the upper and the lower hemiretina have different spatiotemporal response properties; the upper retina is more efficient with stereoscopic stimuli nearer than the fixation point, the lower retina more sensitive for depth stimuli which are farther away. These findings can be interpreted in an ecological framework. Some elaborate motor functions like tool handling rely on the processing of visual information in the closer lower visual field, while more distant visual stimuli occur in the upper visual field. In addition, Murray et al. (1983) reported that the detection of movement measured at 20° eccentricity was better in the lower than in the upper visual field (Fig. 4 of Murray et al. 1983).

In a study on attention and visual search using a "contingent-display technique," Prinz (1984) described differences in detection performance between visual targets presented above or below the fixation point. A vertical asymmetry was found for the detection of an easy target (the letter D) but not for a difficult target (the letter Z). Visual targets below the fixation point could be detected further out in the periphery than targets above the fixation point, and the results were interpreted as an interaction between attention and sensitivity across the visual field. The retinal areas studied were very close to the fovea (about 3° above or below the horizontal meridian), and these data fit well with our results obtained in a simple pattern discrimination experiment where digits presented either 5° above or 5° below the fixation point were significantly better discriminated by the upper retina (see Sect. 5.3).

6.2.2.5 Evoked Potentials

The amplitudes of the evoked potential for stimulation with a spot of 2.5° diameter presented at various points along the horizontal and vertical meridian up to an eccentricity of 30° were described by Copenhaver and Perry (1964), who published data obtained from one trained observer. Evoked potential amplitudes were in general larger for stimuli presented to upper retinal areas at all locations examined except at 5° eccentricity (Fig. 2 of Copenhaver and Perry 1964).

Eason et al. (1967) reported that potentials evoked by flash stimuli have larger amplitudes and shorter latencies when presented to the upper than to the lower hemiretina. These findings were consistent in all three subjects examined.

For checkerboard reversal stimuli, Lehmann and Mir (1976) found a mean latency difference of 23 ms, and an extreme case with a latency difference of 34 ms was reported by Lehmann et al. (1977).

Lehmann and Skrandies (1979) obtained a mean latency difference between upper and lower retinal areas of 14.3 ms for a checkerboard reversal frequency of 1.6 reversals per second, and a mean difference of 11.2 ms for a reversal frequency of 3.2 reversals per second. In addition, for upper hemiretinal stimuli the latencies were significantly shorter for the smaller reversal frequency, but no differences were observed for the lower hemiretina. This indicates that the upper and lower retinal systems may respond independently to a change of stimulus parameters.

Lesèvre (1972) and Lesèvre and Joseph (1979) studied the topography of VEPs using chronotopograms recorded from longitudinal chains of electrodes. For lower hemiretinal stimuli the major component was located occipitally below the inion, and thus the component was missed in some recordings (e. g., Fig. 10 of Lesèvre 1972). When the lower hemiretinal response could be identified, the component latency was delayed by about 10 ms compared to that of the upper retina (Lesèvre and Joseph 1979).

The interpretation of potential waveshapes led Jeffreys and Smith (1979) to the conclusion that the main difference between potentials evoked by upper and lower retinal stimuli was a difference in scalp distribution. Using brief onset/offset stimulus presentations, these authors reported a polarity inversion between upper and lower evoked components. Since the latencies investigated were not identical, and owing to the complexity in interpretation of potential waveshape data (outlined in Sect. 2), one has some reservations as to the interpretation of the data. Lehmann et al. (1977) have shown that polarity inversions may be erroneously concluded when information on the zero voltage baseline is disregarded. The selection of appropriate electrode positions may introduce additional problems for the interpretation of evoked brain activity. More interestingly, Jeffreys and Smith (1979) presented in their Table 1 latency data which show consistently shorter component latencies for upper than lower retinal stimuli. This difference averages 8.3 ms and is in the same order as our latency differences obtained from topographical scalp field data (see Sect. 3).

Statistically significant latency differences between components evoked by stimulation of upper and lower retinal areas were also demonstrated with different presentation modes: Skrandies et al. (1980) found a mean latency difference between upper and lower retinal stimuli of 11 ms for pattern onset, of 8 ms for reversal, and a nonsignificant difference of 3 ms for offset. Further supporting evidence from similar experiments is reported in Sect. 3.2.2.

In a related study, Kriss and Halliday (1980) showed that for lower retinal stimuli a rather broad positive peak was observed around 120 ms while upper retinal stimulation yielded a generalized positivity at 100 ms.

Kimura and Tsutsui (1981) found in ten subjects a mean latency difference of 44 ms for grating stimuli presented either at 15° above or 15° below the fixation point. This difference was statistically significant.

A large subject population was examined by Kuba et al. (1982) using a checkerboard reversal stimulus. Based on the data of 60 subjects, they reported a statistically significant mean latency difference of 17 ms between upper and lower retinal stimulation.

Using steady-state evoked potentials, Yanashima (1982) described larger amplitudes evoked by upper retinal than by lower retinal checkerboard reversal stimuli over the occipital areas. Yanashima (1982) also reported a phase difference of about 180° for the 12 Hz component which corresponds to a latency difference of about 42 ms. The distribution of the electrical activity is hard to interpret, since the electrode array did not extend to below the inion and only the absolute power of the Fourier components was analyzed. When considering the neuronal generating mechanisms (see Fig. 45), it is not surprising that Yanashima (1982) found larger amplitudes with lower retinal stimuli over anterior scalp areas: The power values are polarity-independent, and it appears conceivable that a relative negativity is more pronounced over distal scalp areas [see also Sect. 2 and Lehmann (1984) for a discussion of the scalp mapping of spectral power].

Adachi-Usami and Lehmann (1983) investigated the differences between monocular and binocular evoked potentials and found significant latency differences between upper and lower retinal stimuli in all experimental conditions. For small check sizes (14′ arc), the latency differences averaged 15.2 ms for monocular and 12.3 ms for binocular stimuli; for a check size of 56′ arc, the monocular differences between upper and lower retinal stimuli were 13.2 ms and the binocular differences were 15.5 ms.

Skrandies (1984) reported for vertical grating stimuli of 2.3 c/deg a mean latency difference between components evoked by upper or lower retinal stimulation of 5.4 ms. These data were obtained in a population of 12 healthy adults and the latency differences were highly significant.

All these human evoked potential results are in good agreement with the findings presented in Sect. 3, where shorter component latencies were obtained for scalp

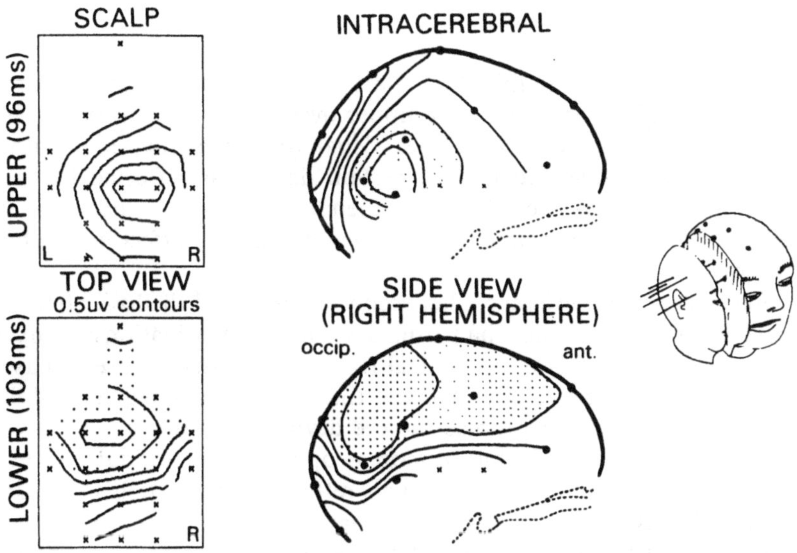

Fig. 45. Intracerebral and scalp potential fields evoked by upper or lower hemiretinal checkerboard reversal stimuli recorded simultaneously in 48 channels (20 scalp and 28 intracranial electrodes). Intracerebral recording sites are marked by *dots*, and these potential fields are located in the parasagittal plane 3 cm from the midline (*right*). (From Lehmann et al. 1982; by permission of Raven Press, New York)

field data evoked by upper retinal stimuli. In addition to the latency differences there are also differences in scalp location of the major evoked components. The general finding that components evoked by upper retinal stimulation are located more anteriorly on the scalp than components evoked by lower hemiretinal stimuli agrees well with the underlying physiologic projection of the visual field onto the visual cortex (see Sect. 6.1). Similar location differences have also been described by Lehmann et al. (1977), Lehmann and Skrandies (1979), Lesèvre and Joseph (1979), Skrandies et al. (1980), Skrandies (1984a,b), or Adachi-Usami and Lehmann (1983).

In a study on intracerebrally recorded potential fields in human patients (Lehmann et al., 1982), we have had the opportunity to study the relationship between intracranial and scalp-recorded brain activity. Figure 45 shows the simultaneously recorded intracerebral and scalp field distributions evoked by upper and lower hemiretinal checkerboard reversal stimuli at the respective component latencies. The intracranial fields in the two stimulus conditions appear to be rotated with respect to each other. According to a stereotactic atlas of the human brain (Talairach and Szikla 1967), the position of the rotation axis corresponds to the middle of the calcarine fissure level. Not only are the positions of the maxima and minima of the evoked fields different, the regions where high potential gradients are located also differ: for upper retinal stimuli the highest potential gradients occur above and for lower retinal stimuli below the level of the calcarine fissure (Fig. 45). These data, in combination with theoretical model dipole computations (Lehmann et al. 1982), illustrate that scalp topography is related to intracranial brain activity in a physiologically meaningful way.

6.3 Clinical Observations

The results presented in this chapter are not intended primarily as further support for functional superiority of the upper over the lower hemiretina. It will rather be shown that the upper and lower visual systems of man are anatomically and physiologically independent to a certain degree, as reflected by the functional results discussed in Sect. 6.2.2. Independence of upper and lower visual fields becomes obvious in various pathological conditions affecting the human visual system.

Some evidence on differences between the upper and the lower retina stems from clinical experience rather than from systematic investigations: Jampolsky (1978) reported that a pathological vertical fixation disparity is observed as an adaptive process of the upper retina only, and that with dissociated vertical divergence in infantile alternating esotropias the eye deviates upward (Jampolsky 1978).

A similar observation is that in most people the eyes are quickly turned upward when the lids are forcibly closed. This reflex, known as Bell's phenomenon, may be regarded as an avoidance reaction that protects the anterior segment of the eye (i. e., the cornea and lens) as well as the upper retina from mechanical damage.

6.3.1 Glaucoma

Both vascular insufficiency and mechanical damage to fibers of the optic nerve have been reported to be caused by elevation of intraocular pressure (Hayreh et al. 1970; Quigley et al. 1980). Electrophysiological studies in the cat have shown that during short-term elevation of intraocular pressure the function of the retinal ganglion cells depends on perfusion pressure and not on the absolute level of intraocular pressure (Grehn and Prost 1983).

Visual field defects in glaucoma show a characteristic pattern, as described in a study on 215 patients by Gramer et al. (1982). Most field defects were found in the upper visual field; this was true for both absolute and relative defects. The authors interpret their findings as a possible indication of a small morphological difference in the routing of nerve fiber bundles of the lower retinal half which may be more susceptible to damage caused by elevated intraocular pressure.

Drance (1977) described the visual fields of 12 eyes of patients with shock-induced optic neuropathy, 28 eyes of patients with low-tension glaucoma, and 35 eyes of patients with chronic open-angle glaucoma. In all three groups the field defects were located more often in the upper than in the lower visual field. In shock-induced neuropathy and low-tension glaucoma upper field defects occurred three times as often as lower field defects, and in chronic glaucoma 64.1% of the field defects were restricted to the lower hemiretina.

When the extent of the visual field is determined by perimetry, some scatter of its boundaries is always observed. Flammer et al. (1984) investigated the factors which influence such short-term fluctuations in automated static perimetry. In patients suffering from glaucoma, larger short-term fluctuations were observed in the upper half of the visual field where field defects also occurred more frequently. A similar tendency appeared in the data of the healthy subjects (see Fig. 3 of Flammer et at. 1984). The main factor for these short-term fluctuations was the differential light sensitivity at the respective test location. This relationship was statistically significant.

6.3.2 Retinal Circulation

Jones and Wilcott (1977) reported impaired dark adaptation and night vision related to physical exercise in four subjects. During exercise the speed of dark adaptation was slower in all subjects, and the threshold for night vision was about 1 log unit higher than during a resting condition. On testing the dark adaptation of peripheral retinal areas (at 10° nasal and up or down) the authors found that in the lower visual fields the decrease in the speed of dark adaptation was even greater than in the upper visual fields, and a markedly elevated threshold for night vision was present in all subjects during the entire exercise interval. The threshold was elevated by 2 log units or more during the first 15 – 20 min, and the elevation remained at a level of about 2 log units during the 60 min-period of exercise. Impaired night vision had not been reported during moderate exercise, and Jones and Wilcott (1977) concluded from their data that the lack of increased ophthalmic arterial pressure might result in a minute paucity of blood in upper retinal areas. This subtle circulatory change as a direct physiological effect may be reflected in the altered dark adaptation.

6.3.3 Epilepsy

Some pattern-sensitive epileptics are specifically responsive to grating patterns of certain orientations. Soso et al. (1980) investigated two pattern-sensitive epileptics stimulated by grating patterns of different orientations presented in the upper or lower visual hemifield. Most EEG spike discharges were found during stimulation of the upper visual field, while the lower visual field yielded only very few spike discharges (see Table 1 of Soso et al. 1980). These findings show that input received via the lower hemiretina has a much higher seizure-facilitating capacity than stimuli presented to the upper hemiretina, supporting the notion that the two visual hemifields may behave as functionally independent systems.

6.3.4 Cerebral Hypoxia

Zihl and von Cramon (1986) described amblyopic visual field defects in patients suffering from chronic cerebral hypoxia. Most of their patients (eight of ten) showed alterations in the lower half of their visual fields. These defects comprised either scotoma or areas with greatly reduced form and color vision. There were, however, no pathophysiological peculiarities in these patients which could account for this asymmetrical distribution of visual field defects (Zihl, personal communication). These observations demonstrate a relative independence of the upper and lower visual hemifields which presumably originates at a cortical level.

7 Summary and General Conclusions

In a series of experiments on human vision I investigated electrophysiological and functional differences between the upper and lower halves of the visual world. The methodological aspect of the work was concerned with adequate biophysical analysis of the electrical brain fields, starting with animal experiments in which volume conduction of brain activity was examined by correlating single unit data with mass activity and going on to the study of the surface-recorded electrical fields of the human brain where multichannel topographical mapping was performed. The aim of these studies was the extraction and statistical analysis of reference-independent descriptors of electrical brain field configurations as recorded from the scalp. Putative neuronal generator sites were described and correlated with earlier results of intracranial evoked potential recordings in human patients.

Since there are fundamental ecological differences between the upper and lower parts of the visual field of man, most of whose objects of interest appear below the horizon, I conducted several series of experiments addressing the question of possible electrophysiological and functional differences in the processing of information in the upper and lower visual fields. The following electrophysiologically ascertained systematic differences support the notion of a functional superiority of the upper hemiretinal system:

1. The component latencies of electrical brain activity evoked by stimuli presented to lower hemiretinal areas were longer than those evoked by upper retinal stimuli, and the scalp topography was consistent with the general neuroanatomical knowledge. These results were found in various subject populations and were independent of the visual targets and the presentation modes: checkerboard and grating stimuli, and pattern onset, reversal, and offset presentations yielded similar differences in brain activity evoked by upper vs lower visual stimuli.
2. The pattern electroretinogram showed similar differences as did the cortical evoked potential with stimulation of upper vs lower retinal areas. These data were obtained in multichannel recordings in which electrical activities originating in the retina and in the visual cortex were separated by topographical methods.
3. The standing potential of the human eye is affected by light stimulation: in electro-oculographic experiments the light response reflecting the interaction between photoreceptors and the retinal pigment epithelium was found to be significantly larger when upper retinal areas were stimulated. In addition, upper retinal areas were more sensitive to changes of the test light intensity.

In psychophysical experiments I was able to demonstrate functional correlates of the electrophysiological findings:

1. Temporal resolution as defined by critical flicker fusion was higher when visual stimuli occurred in the lower visual field. Experiments on double flash discrimination confirmed this finding.
2. Contrast sensitivity was higher for sinusoidal grating patterns presented in the lower than in the upper visual field. These differences were independent of single spatial frequency channels. The conventional visual acuities derived from these data were higher for the upper than the lower retina.

3. In a pattern discrimination task the subjects were better at discriminating stimuli presented on the lower half than on the upper half of a video screen.

Thus, the global superiority of the upper visual system appears to extend from the first step in visual information processing in the retina to higher perceptual and cortical mechanisms, and these functional differences may also have implications for higher visual information processing in conditions involving a man-machine interface.

All differences found were in the same order of magnitude (around 10%), and it is not possible to determine whether the electrophysiological differences observed at later stages (e.g., evoked potential latencies) are directly caused by the differences at earlier stages of the visual system. The data reviewed in Sect. 6 show that anatomical differences between the upper and lower hemiretinal systems exist both at the retinal level (e.g., regional differences in the distribution of photoreceptors) and in the visual cortex, where the lower visual field has a larger representation, and the consistent electrophysiological and functional differences discussed in this chapter may be interpreted as a general system difference.

The differences between the upper and lower hemiretinal systems of man are probably related to the fact that the upper and lower halves of the outer visual world are of different importance for the organism. The function of perception is the acquisition of useful information about the environment. Hence, brain mechanisms of visual information processing must be adapted and optimized to the specific environment of the organism. Information about the outer world in turn enables the animal to respond adequately to sensory stimuli as well as to complex stimulus configurations. Contrary to simplifying laboratory conditions, in which light spots, isolated contours, or near-threshold visual stimuli are presented, in real life the processing of complex patterns of physical stimuli also takes into account the meaningfulness of the stimuli in the outer world. The environmental parameters determine all perceptual mechanisms and capabilities [see also Gibson (1966) for a discussion of such an ecological approach to perception]. Survival requires the ability of the organism to select perceivables on the basis of their importance, and thus the aim of the sensory systems must be to tell the organism what is out there. Sensory information is compared with memory contents in order to achieve adequate interpretations of the environmental information and to adjust the systems' functional state via the orienting response to a condition which optimally permits the selection and execution of adequate reactions to the stimulus configuration.

In addition, the ability to concentrate on certain features of the constant information flow is important. The organism must be able to select useful, relevant information while ignoring other stimuli, and it must also be alert to any changes in the stimulus configurations.

There are pronounced differences between species, even in the gross anatomy of sensory systems' which reflect differences in phylogenesis and habitat. For instance, prey and predator differ: The rabbit's eyes are located laterally on the head, yielding a panoramic field of view. Thus, the animal is able to detect enemies approaching from any direction. For hunting animals the binocular field of view is more important, and fox and wolf have frontal eyes with a smaller total visual field but a relatively large binocular field. Ecological factors and evolution press for optimal adaptation of the sensory systems to the environment (Gibson 1966).

The differences in significance between the upper and lower visual fields fit nicely into such an ecological framework. Stimuli occurring above or below the visual horizon typically carry different meaning and have a different degree of importance for the organism. For most mammals, the upper visual field appears to be less important than the lower. The upper visual field includes mainly vegetation and sky and contains less texture than the lower visual field, most things of interest are located in the lower hemisphere of the visual world. The horizon is the natural optical boundary between sky and ground. Most organisms are capable of detecting verticality, which is also defined by gravity, and they orient according to the difference between up and down.

The basic general difference between the upper and lower visual hemipheres is that direct light always comes from above. Sunlight is very intense, and therefore the lower hemiretina, which is less sensitive than the upper, is well adapted to the light intensity. The convex-concave illusion described in Sect. 6.2.2.4 illustrates that our visual system always assumes that light comes from above.

Light coming from the lower visual field is reflected from object surfaces and is less intense. Prey or enemies of most mammals which do not have birds as predators occur in the lower visual field. For man also, most objects of interest are below the horizon, and the upper visual system must analyse fine textures and the details of the visual world. Therefore, the upper visual system is endowed with higher sensitivity and with better spatial and temporal resolution capacities than the lower visual areas. In addition, the coordination of man's hands in the lower binocular visual field is guided by vision and locomotion relies mainly on information from the lower visual hemisphere, and there are also differences in stereoscopic information processing between the upper and lower visual fields, as reviewed above.

There are meaningful exceptions to this upper visual system superiority in some species: The pigeon retina shows different pigmentation in its upper and lower parts, making the lower retina more sensitive for blue light. In addition, mainly the lower retina of the pigeon is able to detect the plane of polarized light. This enables the animal to orient according to the polarization pattern of the skylight coming from above.

In small animals like the rat, mouse, or frog, evidence has been obtained of asymmetry in midbrain structures of the visual system favoring the upper visual hemisphere. The upper visual field may carry important information for these animals. Enemies of the mouse and the rat may approach from above, and frogs have to catch flies in the upper binocular field of view.

The reviewed human anatomical and clinical data show that the upper and lower retinal halves may be considered as physiologically independent systems. In support of this conjecture, functional differences between upper and lower retinal areas were found in human psychophysical experiments which show that visual acuity, temporal sensitivity, and pattern discrimination are better with stimuli presented in the lower than in the upper visual field.

The data obtained in electrophysiological experiments correlate well with the observed functional differences between the upper and lower visual systems. The electrical fields of the human brain recorded from the intact scalp were shown to be directly related to the underlying neuronal processes, and topographical analyses of these potential fields demonstrated significant differences between the processing of upper and lower retinal stimulus material. Consistent differences in scalp topography and in component latency of the evoked components are found in various stimulus conditions and subject populations.

In addition, the differences between upper and lower areas can already be demonstrated at the level of the retina. The functional superiority of the upper visual system thus appears to extend from the first step in visual information processing in the retina to higher, complex perceptual mechanisms, as reflected by the better performance in pattern discrimination tasks where meaningful stimulus material is used.

Moreover, during ontogenetic development there is an interaction between the functional aspects of information processing and the organization of the sensory systems. In cats it has been shown that early visual experience alters the physiological responses of the visual system. As a consequence of early light deprivation during a sensitive period, the number of cortical neurons that later in life can be influenced by light declines (Hubel and Wiesel 1970b). Hirsch and Spinelli (1970) studied cats which were raised from birth with one eye viewing vertical lines and the other eye viewing horizontal lines, and they found that the receptive fields of cortical neurons were predominantly oriented according to this early experience. These studies illustrate that functional neuronal connections can be selectively modified by environmental stimulus configurations. Similarly, Annis and Frost (1973) described orientation anisotropies in human visual acuity of adult subjects. These anisotropies appeared to be directly related to the early visual environment in which the subjects had been raised.

Perceptual capability is determined phylogenetically and ontogenetically both by anatomical factors and by the structure of the environment. Thus, sensory systems are optimally adapted to the habitat, and deal with the demands of specific aspects of visual information processing. In the case of the division of the outer world into an upper and a lower half, it can be argued that phylogenetically, human prey and enemies occurred mainly in the lower parts of the visual field, and therefore improved performance of the human upper visual system was advantageous.

Acknowledgements. The experiments presented in this text were performed in the course of the past few years, and several persons contributed to different aspects of this final summary. First of all, I wish to thank Professor Dr. E. Dodt for his continuing support of my work and Professor Dr. D. Lehmann for many fruitful and inspiring discussions and his close cooperation on the topography of electrical brain activity. Professor Dr. H. Wässle and Dr. L. Peichl gave useful hints on anatomical findings in the retina. Professor Dr. C. Baumann, Professor Dr. E. Dodt, Professor Dr. D. Lehmann, and Dr. D. Brandeis read a first draft of the manuscript and made valuable suggestions for this final version. Thanks are also due to Dr. A. Raile, who helped with the data collection in some of the experiments described in chapter 4.2; to Dr. O. Ludwig for his advice concerning statistical problems; and to Mrs. H. Schneider, who prepared most of the figures.

8 References

Adachi-Usami E, Lehmann D (1983) Monocular and binocular evoked average potential field topography: upper and lower hemiretinal stimuli. Exp Brain Res 50:341–346

Aebersold H, Creutzfeldt OD, Kuhnt U and Sanides D (1981) Representation of the visual field in the optic tract and optic chiasm of the cat. Exp Brain Res 42:127–145

Afanador AJ, Andrews CE (1978) Rod and cone contribution to the EOG ratio. Am J Optom 55:101–107

Ali MA, Klyne MA (1985) Vision in vertebrates. Plenum New York

Alpern M (1972) Eye movements. In: Jameson D, Hurwich LM (eds) Handbook of sensory physiology, vol VII/4. Springer, Berlin Heidelberg New York, pp 304–330

Annis RC, Frost B (1973) Human visual ecology and orientation anisotropies in acuity. Science 182:729–731

Arden GB, Kelsey JH (1962a) Changes produced by light in the standing potential of the human eye. J Physiol 161, 189–204

Arden GB, Kelsey JH (1962b) Some observations on the relationship between the standing potential of the human eye and the bleaching and regeneration of visual purple. J Physiol 161, 205–226

Arden GB, Carter RM, Hogg C, Siegel IM, Margolis S (1979) A gold foil electrode: extending the horizons for clinical electroretinography. Invest Ophthalmol 18, 421–426

Armington JC (1968) The electroretinogram, the visual evoked potential, and the area-luminance relation. Vision Res 1968, 8, 263–276

Armington JC (1974) The electroretinogram. Academic, New York

Aschoff U (1981) Skotopische und photopische Anteile der Hell- und Dunkelschwingung im Elektrookulogramm. Dev Ophthalmol 4, 149–166

Aubert H, Förster R (1857) Beiträge zur Kenntnis des indirecten Sehens. I. Untersuchungen über den Raumsinn der Retina. Arch Ophthalmol 3, 1–37

Baizer JS, Maguire WM (1983) Double representation of lower visual quadrant in prelunate gyrus of rhesus monkey. Invest Ophthalmol Vis Sci 24, 1436–1439

Bartlett NR, Sticht TG, Pease VP (1968) Effects of wavelength and retinal locus on the reaction time to onset and offset stimulation. J Exp Psychol 78, 699–701

Basler A (1911) Über die Verschmelzung von zwei nacheinander erfolgenden Lichtreizen. Pflügers Arch 143, 245–251

Bennet-Clark HC (1964) The oculomotor response to small target replacements. Optica Acta 11, 301–314

Berger H (1929) Über das Elektrenkephalogramm des Menschen. 1. Mitteilung. Arch Psychiat Nervenkrankh 87, 527–570

Bilge M, Bingle A, Seneviratne KG, Whitteridge D (1967) A map of the visual cortex in the cat. J Physiol 191, 116P–118P

Bishop PO, Kozak W, Vakkur GJ (1962) Some quantitative aspects of the cat's eye: axis and plane of reference, visual field co-ordinates and optics. J Physiol 163, 466–502

Bjaalie JG (1985) Distribution in areas 18 and 19 of neurons projecting to the pontine nuclei: a quantitative study in the cat with retrograde transport of HRP-WGA. Exp Brain Res 57, 585–597

Bjaalie JG, Brodal P (1983) Distribution in area 17 of neurons projecting to the pontine nuclei: a quantitative study in the cat with retrograde transport of HRP-WGA. J Comp Neurol 221:289–303

Bodis-Wollner I, Diamond SP (1976) The measurement of spatial contrast sensitivity in cases of blurred vision associated with cortical lesions. Brain 99, 695–710

Braddick D, Campbell FW, Atkinson J (1978) Channels in vision: basic aspects. In: Held R, Leibowitz HW, Teuber H-J (eds) Handbook of sensory physiology, vol 8. Springer, Berlin Heidelberg New York, pp 3–38

Breitmeyer B, Julesz B, Kropfl W (1975) Dynamic random-dot stereograms reveal up-down anisotropy and left-right isotropy between cortical hemifields. Science 187, 269–270
Brettel H, Caelli T, Hilz R, Rentschler I (1982) Modelling perceptual distortion: amplitude and phase transmission in the human visual system. Hum Neurobiol 1, 61–67
Brindley G, Lewin WS (1968) The sensations produced by electrical stimulation of the visual cortex. J Physiol 196, 479–493
Broca P (1861) Perte de la parole. Romolissement chronique et destruction partielle du lobe anterieur gauche du cerveau. Bull Soc Anthrop (Paris) 219
Brown JL (1965) Flicker and intermittent stimulation. In: Graham CH (ed) Vision and visual perception. Wiley, New York, pp 251–320
Burkhalter A, Felleman DJ, Newsome WT, van Essen DC (1986) Anatomical and physiological asymmetries related to visual areas V3 and VP in macaque extrastriate cortex. Vision Res 26, 63–80
Campbell FW, Robson JG (1968) Applications of Fourier analysis to the visibility of gratings. J Physiol 197, 551–566
Caton R (1875) The electric currents of the brain. Br Med J 2, 278
Clarke PGH, Whitteridge D (1976) The projection of the retina, including the red area, on to the optic tectum of the pigeon. Q J Exp Physiol 61, 351–358
Cocito L, Favale E, Tartaglione A (1977) Asimmetrie funzionali tra emicampo visivo superiore ed inferiore nel soggetto normale. Boll Soc It Biol Sper 53, 629–633
Copenhaver RM, Perry NW (1964) Factors affecting visually evoked cortical potentials such as impaired vision of varying etiology. Invest Ophthalmol 3, 665–675
Creutzfeldt OD (1983) Cortex cerebri. Springer Berlin Heidelberg New York
Creutzfeldt OD, Kuhnt U (1967) The visual evoked potential: physiological, developmental, and clinical aspects. Electroencephalogr Clin Neurophysiol [Suppl 26]:29–41
Curcio CA, Hendrickson AE, Kalina RE (1985) Topographical distribution of human photoreceptors. Invest Ophthalmol Vis Sci 26 [Suppl]:261
Dawson WW, Maida TM (1984) Relations between the human retinal cone and ganglion cell distribution. Ophthalmologica 188, 216–221
Delius JD, Perchard RJ, Emmerton J (1976) Polarized light discrimination by pigeons and an electroretinographic correlate. J Comp Physiol Psychol 90, 560–571
Dimond SJ, Beaumont JG (1974) Hemisphere function in the human brain. Elek Science, London, p 398
Ditchburn RW (1973) Eye movements and visual perception. Clarendon, Oxford
Dodt E (1951) Cone electroretinography by flicker. Nature 168:738
Dodt E (1964) Erregung und Hemmung retinaler Neurone bei intermittierender Belichtung. Doc Ophthalmol 18, 259–274
Dodt E, Baier M (1984) Area-luminance relationship for a constant light peak of the standing potential in the human eye. Ophthalmologica 188, 232–238
Dodt E, Enroth C (1954) Retinal flicker response in cat. Acta Phys Scand 30, 375–390
Donchin E, Ritter W, McCallum WC (1978) Cognitive psychophysiology: the endogenous components of the ERP. In: Callaway E, Tueting P, Koslow SH (eds), Event-related brain potentials in man. Academic, New York, pp 349–411
Dräger UC, Hubel DH (1976) Topography of visual and somatosensory projections to mouse superior colliculus. J Neurophysiol 39, 91–101
Drance SM (1977) The visual field of low tension glaucoma and shock-induced optic neuropathy. Arch Ophthalmol 95, 1359–1361
Drasdo N (1977) The neural representation of visual space. Nature 266, 554–555
Eason RG, White CT, Oden D (1967) Averaged occipital responses to stimulation of sites in the upper and lower halves of the retina. Percept Psychophys 2, 423–425
Ehrlich D (1981) Regional specialization of the chick retina as revealed by the size and density of neurons in the ganglion cell layer. J Comp Neurol 195, 643–657

Elenius V, Aantaa E (1973) Light-induced increase in amplitude of electro-oculogram. Arch Ophthalmol 90, 60–63

Emmerton J (1983a) Functional morphology of the visual system. In: Abs M (ed), Physiology and behaviour of the pigeon, Academic, New York, pp 221–244

Emmerton J (1983b) Vision. In: Abs M (ed), Physiology and behaviour of the pigeon, Academic, New York, pp 245–266

Estevez O, Spekreijse H (1974) Relationship between pattern appearance - disappearance and pattern reversal response. Exp Brain Res 19, 233–238

Flammer J, Drance SM, Fankhausen F, Augustiny L (1984) Differential light threshold in automated static perimetry. Arch Ophthalmol 102, 876–879

Fox SS, O'Brien JH (1965) Duplication of evoked potential waveform by curve of probability of firing of a single cell. Science 147, 888–890

Freeman RB (1964) Figural after-effects: displacement or contrast. Am J Pyschol 77, 607–613

Freeman WJ (1978) Discussion in E. Donchin: Use of scalp distribution as a dependent variable in event-related potential studies: excerpts of preconference correspondence. In: Otto DA (ed) Multidisciplinary perspectives in event-related brain potential research. EPA, Washington, pp 501–510

Gazzaniga MS (1970) The bisected brain. Appleton-Century-Crofts, New York

Gibson A, Baker J, Mower G, Glickstein M (1978) Corticopontine cells in area 18 of the cat. J Neurophysiol 41, 484–495

Gibson JJ (1966) The senses considered as perceptual systems. Houghton Mifflin, Boston

Glickstein M, Stein J, King RA (1972) Visual input to the pontine nuclei. Science 178, 1110–1111

Goodale MA (1983) Visually guided, pecking in the pigeon (*Columba livia*). Brain Behav Evol 22, 22–41

Gramer E, Gerlach R, Krieglstein GK, Leydhecker W (1982) Zur Topographie früher glaukomatöser Gesichtsfeldausfälle bei der Computerperimetrie. Klin Monatsbl Augenheilk 180, 515–523

Granit R (1947) Sensory mechanisms of the retina. Hafner, New York

Granit R (1955) Centrifugal and antidromic effects on ganglion cells of retina. J Neurophysiol 18, 388–411

Granit R, Hammond EL (1931) Comparative studies on the peripheral and central retina. V. The sensation-time curve and time course of the fusion frequency on intermittent stimulation. Am J Physiol 98, 654–663

Greenberg JH, Reivich M, Alavi A, Hand P, Rosenquist A, Rintelmann W, Stein A, Tusa R, Dann R, Christman D, Fowler J, McGregor B, Wolf A (1981) Metabolic mapping of functional activity in human subjects with the [18-F]fluorodeoxyglucose technique. Science 212, 678–680

Grehn F, Prost M (1983) Function of retinal nerve fibers depends on perfusion pressure: neurophysiologic investigations during acute intraocular pressure elevation. Invest Ophthalmol 24, 347–353

Griff ER, Steinberg RH (1982) Origin of the light peak: in vitro study of *Gekko gekko*. J Physiol 331, 637–652

Groneberg A, Teping C (1980) Topodiagnostik von Sehstörungen durch Ableitung retinaler und kortikaler Antworten auf Umkehr-Kontrastmuster. Ber Dtsch Ophthalmol Ges 77, 409–415

Grüsser O-J (1984) Face recognition within the reach of neurobiology and beyond it. Hum Neurobiol 3, 183–190

Grüsser O-J, Kapp H (1958) Reaktionen retinaler Neurone nach Lichtblitzen. II. Doppelblitze mit wechselndem Blitzintervall. Pflügers Arch 266, 111–129

Grüsser O-J, Rabelo C (1958) Reaktionen retinaler Neurone nach Lichtblitzen. I. Einzelblitze und Blitzreize wechselnder Frequenz. Pflügers Arch 265, 501–525

Gstalder RJ, Green DG (1971) Laser interferometric acuity in amblyopia. J Pediatr Ophthalmol 8, 251–256

Hall GS, von Kries J (1879) Über die Abhängigkeit der Reaktionszeit vom Ort des Reizes. Arch Anat Physiol (Leipzig) [Suppl] 1–10

Halliday AM, McDonald WI, Mushin J (1973) The visual evoked response in the diagnosis of multiple sclerosis. Br Med J 4, 661–664

Hayreh SS, Revie IHS, Edwards J (1970) Vasogenic origin of visual field defects and optic nerve changes in glaucoma. Br J Ophthalmol 54, 461–472

Hebel R, Holländer H (1983) Size and distribution of ganglion cells in the human retina. Anat Embryol 168, 125–136

Hirsch HVB, Spinelli DN (1970) Visual experience modifies distribution of horizontally and vertically oriented receptive fields in cats. Science 168, 869–871

Hjorth B (1975) An on-line transformation of EEG scalp potentials into orthogonal source derivations. Electroencephalogr Clin Neurophysiol 39, 526–530

Holden AL, Powell TPS (1972) The functional organization of the isthmo-optic nucleus in the pigeon. J Physiol 223, 419–447

Holländer H, Bisti S, Maffei L, Hebel R (1984) Electroretinographic responses and retrograde changes of retinal morphology after intracranial optic nerve section. A quantitative analysis in the cat. Exp Brain Res 55, 483–493

Holm S (1979) A simple sequentially rejective multiple test procedure. Scand J Statist 6, 65–70

Holmes G (1945) The organization of the visual cortex in man. Proc R Soc, Lond (Biol) 132, 348–361

Hubel DH, Wiesel TN (1970a) Cells sensitive to binocular depth in area 18 of the macaque monkey cortex, Nature 225, 41–42

Hubel DH, Wiesel TN (1970b) The period of susceptibility to the physiological effects of unilateral eye closure in kittens. J Physiol 206, 419–436

Hughes A (1975) A quantitative analysis of the cat retinal ganglion cell topography. J Comp Neurol 163, 107–128

Hughes A, Wässle H (1976) The cat optic nerve: fibre total count and diameter spectrum. J Comp Neurol 169, 171–184

Hylkema BS (1942) Examination of the visual field by determining the fusion frequency. Acta Ophthalmol 20, 181–193

Jampolsky A (1978) Unequal visual inputs in strabismus management: a comparison of human and animal strabismus. In: Symposium on strabismus. Transactions of the New Orleans Academy of Ophthalmology. Mosby, St. Louis, pp 358–492

Jeeves MA (1984) The historical roots and recurring issues of neurobiological studies of face perception. Hum Neurobiol 3, 191–196

Jeffreys DA, Smith AT (1979) The polarity inversion of scalp potentials evoked by upper and lower half field stimulus patterns: latency or surface distribution differences? Electroencephalogr Clin Neurophysiol 46, 409–415

Jewett DL, Romano MN, Williston JS (1970) Human auditory evoked potentials: possible brain stem components detected on the scalp. Science 167, 1517–1518

Jones RK, Wilcott IT (1977) Topographic impairment of night vision related to exercise. Am J Ophthalmol 84, 868–871

Julesz B (1971) Foundations of cyclopean perception. University of Chicago Press, Chicago

Kavanagh RN, Darcey TM, Lehmann D, Fender DH (1978) Evaluation of methods for three-dimensional localization of electrical sources in the human brain. IEEE Trans Biomed Eng 25, 421–429

Kimura H, Tsutsui J (1981) Average responses evoked by moving grating pattern in the upper, central and lower visual field. Neurosci Lett 24, 295–299

King-Smith PE (1969) Absorption spectra and function of the colored oil drops in the pigeon retina. Vision Res 9, 1391–1399

Kleberger E (1955) Untersuchungen über die Verschmelzungsfrequenz intermittierenden Lichts an gesunden und kranken Augen. III. Das normale Flimmergesichtsfeld. Graefes Arch Ophthalmol 157, 158–166

Kriss A, Halliday AM (1980) A comparison of occipital potentials evoked by pattern onset, offset and reversal by movement. In: Barber C (ed) Evoked potentials. MTP Press, Lancaster, pp 205–212

Kuba M, Peregrin K, Vit F, Hanusova I (1982) Visual evoked responses to reversal stimulation in the upper and lower half of the central part of the visual field in man. Physiol Bohemeslow 31, 503–510

Landau D, Dawson WW (1970) The histology of retinas from the pinnipedia. Vision Res 10, 691–702

Landis C (1954) Determinants of critical flicker-fusion threshold. Physiol Rev 34, 259–286

Landolt E, Hummelsheim E (1904) Die Untersuchung der Funktionen des excentrischen Netzhautgebietes. In: Sämisch T (ed) Gräfe-Sämisch Handbuch der gesamten Augenheilkunde, vol 4. Engelmann, Leipzig, pp 503–583

Lawden MC (1982) The analysis of spatial phase in amblyopia. Hum Neurobiol 1, 55–60

Lehmann D (1977) The EEG as scalp field distribution. In: Rémond A (ed) EEG informatics. Elsevier, Amsterdam, pp 365–384

Lehmann D (1984) EEG measurement of brain activity: spatial aspects, segmentation and imaging. Int J Psychophysiol 1, 267–276

Lehmann D, Julesz B (1978) Lateralized cortical potentials elicited by dynamic random dot stereograms. Vision Res 18, 1265–1271

Lehmann D, Mir Z (1976) Methodik und Auswertung visuell evozierter EEG-Potentiale bei Verdacht auf multiple Sklerose. J Neurol 213, 97–103

Lehmann D, Skrandies W (1979) Multichannel evoked potential fields show different properties of human upper and lower hemi-retinal systems. Exp Brain Res 35, 151–159

Lehmann D, Skrandies W (1980) Reference-free identification of components of checkerboard-evoked multichannel potential fields. Electroencephalogr Clin Neurophysiol 48, 609–621

Lehmann D, Skrandies W (1984) Spatial analysis of evoked potentials in man: a review. Progr Neurobiol 23, 227–250

Lehmann D, Meles HP, Mir Z (1977) Average multichannel EEG potential fields evoked from upper und lower hemiretina: latency differences. Electroencephalogr Clin Neurphysiol 43, 725–731

Lehmann D, Darcey TM, Skrandies W (1982) Intracerebral and scalp fields evoked by hemiretinal checkerboard reversal, and modeling of their dipole generators. In: Courjon J, Maugiere F, Revol M (eds) Clinical applications of evoked potentials in neurology. Raven, New York, pp 41–48

Lesèvre N (1972) Potentiels évoqués par des patterns chez l'homme: influence des variables caracterisant le stimulus et sa position dans le champ visuel. In: Fessard A, Lelord G (eds) Activités evoquées et leur conditionnement. INSERM, Paris, pp 1–22

Lesèvre N, Joseph JP (1979) Modifications of the pattern-evoked potential (PEP) in relation to the stimulated part of the visual field. Electroencephalogr Clin Neurophysiol 47, 183–203

Lichtenstein M, White CT (1961) Relative visual latency as a function of retinal locus. J Opt Soc Am 51, 1033–1034

Lindsley DB, Lansing RW (1956) Flicker and two-flash fusional threshold and EEG, Am Psychol 11, 433

Linsenmeyer RA, Steinberg RH (1982) Origin and sensitivity of the light peak in the intact cat eye. J Physiol 331, 653–673

Lipkin BS (1962) Monocular flicker discrimination as a function of the luminance and area of contralateral steady light: I. Luminance. II. Area. J Opt Soc Am 52:1287–1295, 1296–1300

Low FN (1943) The peripheral visual acuity of 100 subjects. Am J Physiol 140, 83–88

Lundh BL, Lennerstrand G, Derefeldt G (1983) Central and peripheral normal contrast sensitivity for static and dynamic sinusoidal gratings. Acta Ophthalmol 61, 171–182

Luria SM, Kinney JA (1970) Underwater vision. Science 167, 1454–1461

MacKay DM (1984a) Source density analysis of scalp potentials during evaluated action. I. Coronal distribution. Exp Brain Res 54, 73–85

MacKay DM (1984b) Source density analysis of scalp potentials during evaluated action. II. Lateral coronal distributions. Exp Brain Res 54, 86–94

Maffei L (1982) Electroretinographic and visual cortical potentials in response to alternating gratings. Ann NY Acad Sci 388, 1–10

Maffei L, Fiorentini A (1981) Electroretinographic responses to alternating gratings before and after section of the optic nerve. Science 211, 953–955

Maffei L, Fiorentini A, Bisti S, Holländer H (1985) Pattern ERG in the monkey after section of the optic nerve. Exp Brain Res 59, 423–425

Magrotti E, Cosi V, Borutti G (1980) Differenze funzionali tra emicampi visivi superiore ed inferiore per stimoli non strutturanti. Boll Soc Ital Biol Sper 56, 416–422

Martin GR, Muntz WRA (1979) Retinal oil droplets and vision in the pigeon (*Columba livia*). In: Granda AM, Maxwell JF (eds) Neural mechanisms of behavior in the pigeon. Plenum, New York, pp 307–325

Matelli M, Olivieri MF, Saccani A, Rizzoloatti G (1983) Upper visual space neglect and motor deficits after section of the midbrain commissures in the cat. Behav Brain Res 10, 263–285

McAlpine D, Lumsden CE, Acheson ED (1972) Multiple sclerosis: a reappraisal. Williams and Wilkins, Baltimore

Millodot M, Lamont A (1974) Peripheral visual acuity in the vertical plane. Vision Res 14, 1497–1498

Mitzdorf U (1985) Current source-density method and application in cat cerebral cortex: investigation of evoked potentials and EEG phenomena. Physiol Rev 65, 37–100

Mitzdorf U, Singer W (1978) Prominent excitatory pathways in the cat visual cortex (A 17 and A 18): a current source density analysis of electrically evoked potentials. Exp Brain Res 33, 371–394

Murray I, MacCana F, Kulikowski JJ (1983) Contribution of two movement detecting mechanisms to central and peripheral vision. Vision Res 23, 151–159

Nachmias J (1959) Two-dimensional motion of the retinal image during monocular fixation. J Opt Soc Am 49, 901–908

Nagata T, Hayashi Y (1984) The visual field representation of the rat ventral lateral geniculate nucleus. J Comp Neurol 227, 582–588

Newman RP, Kinkel WR, Jacobs L (1984) Altitudinal hemianopia caused by occipital infarctions. Arch Neurol 41, 413–418

Niemeyer G (1975) The function of the retina in the perfused eye. Doc Ophthalmol 35, 53–116

Nuboer JFW, Wortel JF (1985) Wavelength discrimination in the lower and upper visual field of the pigeon. J Physiol 366, 95P

Nunez P (1981) Electric fields of the brain. Oxford University Press, New York

Østerberg G (1935) Topography of the layer of rods and cones in the human retina. Acta Ophthalmol [13 Suppl. 6]: 1–102

Oyster CW, Takahashi ES, Cilluffo M, Brecha NC (1985) Morphology and distribution of tyrosine hydroxylase-like immunoreactive neurons in the cat retina. Proc Natl Acad Sci 82, 6335–6339

Payne WH (1965) Visual reaction times on a circle about the fovea. Science 155, 481–482

Pease VP, Sticht TG (1965) Reaction time as a function of onset and offset stimulation of the fovea and periphery. Percept Mot Skills 20, 549–554
Phelps ME, Kuhl DE, Mazziotta JC (1981) Metabolic mapping of the brain's response to visual stimulation: studies in humans. Science 211, 1445–1448
Phillips G (1933) Perception of flicker in lesions of the visual pathways. Brain 56, 464–478
Poffenberger AT (1912) Reaction time to retinal stimulation with special reference to the time lost in conduction through nerve centers. Arch Psychol 3, 1–73
Poggio GF, Fischer B (1977) Binocular interaction and depth sensitivity in striate and prestriate cortex of behaving rhesus monkey. J Neurophysiol 40, 1392–1405
Presson J, Moran J, Gordon B (1983) Effects of eye rotation on visually guided behavior. J Neurophysiol 50, 631–643
Prinz W (1984) Attention and sensitivity in visual search. Psychol Res 45, 355–366
Quingley HA, Flower RW, Addicks EM, McLeod DS (1980) The mechanism of optic nerve damage in experimental acute intraocular pressure elevation. Invest Ophthalmol 19, 505–517
Rains JD (1963) Signal luminance and position effects in human reaction time. Vision Res 3, 239–251
Regan D, Silver R, Murray TJ (1977) Visual acuity and contrast sensitivity in multiple sclerosis – hidden visual loss. Brain 100, 563–579
Rémond A, Lesèvre N, Joseph JP, Rieger H, Lairy GC (1969) The alpha average: I. Methodology and description. Electroencephalogr Clin Neurophysiol 26, 245–265
Reuter TE, White RH, Wald G (1971) Rhodopsin and porphyropsin in the adult bullfrog retina. J Gen Physiol 58, 351–371
Riemslag FCC, Ringo JL, Spekreijse H, Verduyn Lunel HF (1985) The luminance origin of the pattern electroretinogram in man. J Physiol 363, 191–209
Riggs LA, Johnson EP, Schick AML (1964) Electrical responses of the human eye to moving stimulus patterns. Science 144, 567
Scalia F (1976) The optic pathway of the frog: nuclear organization and connections. In: Llinas R, Precht W (eds) Frog neurobiology. Springer Berlin Heidelberg New York, pp 386–406
Schade OH (1956) Optical and photoelectric analog of the eye. J Opt Soc Am 46, 721–739
Schmidt M, Wässle H, Humphrey M (1985) Number and distribution of putative cholinergic neurons in the cat retina. Neurosci Lett 59, 235–240
Schneider MR (1972) A multistage process for computing dipolar sources of EEG discharges from surface information. IEEE Trans Biomed Eng 19, 1–12
Schneider MR (1974) Effect of inhomogeneities on surface signals coming from a cerebral dipole source. IEEE Trans Biomed Eng 21, 52–54
Schwartz EL, Christman DR, Wolf AP (1984) Human primary visual cortex topography imaged via positron tomography. Brain Res 294, 225–230
Seiple WH, Siegel IM (1983) Recording the pattern electroretinogram: a cautionary note. Invest Ophthalmol 24, 796–798
Seneviratne KN (1963) The representation of the visual field on the subcortical centers of the cat and rabbit. PhD thesis, Edinburgh
Seneviratne KN, Whitteridge D (1962) Visual evoked responses in the lateral geniculate nucleus. Electrocephalogr Clin Neurophysiol 14, 785
Shickman GM (1981) Time-dependent functions in vision. In: Moses RA (ed) Adler's physiology of the eye. Mosby, St. Louis, pp 663–713
Shurong W, Kun Y, Yinting W (1981) Visual field topography and binocular responses in frog's nucleus isthmi. Scientia Sin [B] 24, 1292–1301
Sidman RD, Giambalvo V, Allison T, Bergey P (1978) A method for localization of sources of human cerebral potentials evoked by sensory stimuli. Sensory Proc 2, 116–129
Simonson E (1958) Contralateral glare effect on the fusion frequency of flicker. Arch Ophthalmol 60, 995–999

Skrandies W (1981) Latent components of potentials evoked by visual stimuli in different retinal locations. Int J Neurosci 14, 77–84

Skrandies W (1983) Information processing and evoked potentials: topography of early and late components. Adv Biol Psychiatr 13, 1–12

Skrandies W (1984a) Differences of visual evoked potential latencies and topographies depending on retinal location and presentation mode. Pflügers Arch 400, R31

Skrandies W (1984b) Scalp potential fields evoked by grating stimuli: effects of spatial frequency and orientation. Electroencephalogr Clin Neurophysiol 58, 325–332

Skrandies W (1985a) Critical flicker fusion and double flash discrimination in different parts of the visual field. Int J Neurosci 25, 225–231

Skrandies W (1985b) Human contrast sensitivity: regional retinal differences. Hum Neurobiol 4, 95–97

Skrandies W (1986a) Visual evoked potential topography: methods and results. In: Duffy FH (ed) Topographic mapping of brain electrical activity. Butterworth, Boston, pp 7–28

Skrandies W (1986b) Temporal summation of stereoscopic visual stimuli: brain electric components and subjective perception. In: Rohrbaugh JW, Johnson R, Parasuraman R (eds) Research reports of EPIC VIII conference, Stanford, pp 397–399

Skrandies W, Baier M (1986) Activity of human pigment epithelium shows differences between upper and lower retinal areas. Vision Res 26, 577–581

Skrandies W, Gottlob I (1986) Alterations of visual contrast sensitivity in Parkinson's Disease. Hum Neurobiol 5, 255–259

Skrandies W, Lehmann D (1982a) Occurrence time and scalp location of components of evoked EEG potential fields. In: Herrmann WM (ed) Electroencephalography in drug research. Fischer, Stuttgart, pp 183–192

Skrandies W, Lehmann D (1982b) Spatial principal components of multichannel maps evoked by lateral visual half-field stimuli. Electroencephalogr Clin Neurophysiol 54, 662–667

Skrandies W, Vomberg HE (1985) Stereoscopic stimuli activate different cortical neurones in man: electrophysiological evidence. Int J Psychophysiol 2, 293–296

Skrandies W, Richter M, Lehmann D (1980) Checkerboard evoked potentials: topography and latency for onset, offset, and reversal. Progr Brain Res 54, 291–295

Skrandies W, Wässle H, Peichl L (1978) Are field potentials an appropriate method for demonstrating connections in the brain? Exp Neurol 60, 509–521

Skrandies W, Chapman RM, McCrary JW, Chapman JA (1984) Distribution of latent components related to information processing. Ann NY Acad Sci 425, 271–277

Soso MJ, Lettich E, Belgum JH (1980) Pattern-sensitive epilepsy. II: Effects of pattern orientation and hemifield stimulation. Epilepsia 21, 313–323

Southall JPC (1962) Helmholtz's treatise on physiological optics. Dover, New York

Spalding JMK (1952) Wounds of the visual pathway. J Neurol Neurosurg Psychiatr 15, 169–183

Sperry RW (1964) The great cerebral commissure. Sci Am 210, 42–52

Sperry RW (1968) Hemisphere deconnection and unity in conscious awareness. Am Psychol 23, 723–733

Standage GP, Benevento LA, The organization of connections between the pulvinar and visual area MT in the macaque monkey. Brain Res 262, 288–294

Starr A, Achor J (1975) Auditory brainstem responses in neurological disease. Arch Neurol 32, 761–768

Stone J, Johnston E (1981) The topography of primate retina: a study of the human, bushbaby, and New- and Old-World monkeys. J Comp Neurol 196, 205–223

Sutton S, Braren M, Zubin J, John ER (1965) Evoked potential correlates of stimulus uncertainty. Science 150, 1187–1188

Talairach J, Szikla G (1967) Atlas of stereotaxic anatomy of the telencephalon. Masson, Paris

Täumer R (ed) (1976) Electro-oculography – its clinical importance. Karger, Basel. Bibliotheca Ophthalmologica, Vol 85
ten Doesschate J (1946) Visual acuity and distribution of percipient elements on the retina. Ophthalmologica 112, 1–18
Teuber H-L, Battersby WS, Bender MB (1960) Visual field defects after penetrating missile wounds of the brain. Harvard University Press, Cambridge
Teuber ML (1974) Sources of ambiguity in the prints of Maurits C. Escher. Scie. Am. 231 (No 1), 90–104
Torrealba F, Guillery RW, Eysel U, Polley EH, Mason CA (1982) Studies of retinal representations within the cat's optic tract. J Comp Neurol 211, 377–396
Tusa RJ, Palmer LA, Rosenquist AC (1978) The retinotopic organization of area 17 (striate cortex) in the cat. J Comp Neurol 177, 213–236
Valeton JM, van Norren D (1982) Intraretinal recording of slow electrical responses to steady illumination in monkey: isolation of receptor responses and the origin of the light peak. Vision Res 22, 393–399
van Buren A (1963) The retinal ganglion cell layer. Thomas, Springfield
van de Grind WA, Grüsser O-J, Lunkenheimer H-U (1973) Temporal transfer properties of the afferent visual system. Psychophysical, neurophysiological and theoretical investigations. In: Jung R (ed) Handbook of sensory physiology, Vol VII/3A. Springer, Berlin Heidelberg New York, pp 431–573
van der Waerden BL (1971) Mathematische Statistik, 3rd ed. Springer, Berlin Heidelberg New York
van Essen DC (1985) Functional organization of primate visual cortex. In: Peters A, Jones EG (eds) Cerebral cortex, Vol 3. Plenum, New York, pp 259–329
van Essen DC, Maunsell JHR, Bixby JL (1981) The middle temporal visual area in the macaque: myeloarchitecture, connections, functional properties and topographic organization. J Comp Neurol 199, 293–326
van Essen DC, Newsome WT, Maunsell JHR (1984) The visual field representation in striate cortex of the macaque monkey: asymmetries, anisotropies, and individual variability. Vision Res 24, 429–448
van Essen DC, Newsome WT, Maunsell JHR, Bixby JL (1986) The projections from striate cortex (V1) to areas V2 and V3 in the macaque monkey: asymmetries, areal boundaries, and patchy connections. J Comp Neurol 244, 451–480
Vaney DI (1985) The morphology and topographic distribution of AII amacrine cells in the cat retina. Proc R Soc Lond [Biol] 224, 475–488
Vaney DI, Hughes A (1976) Rabbit optic nerve: fibre diameter spectrum, fibre count, and comparison with a retinal ganglion cell count. J Comp Neurol 170, 241–251
Vomberg HE, Skrandies W (1985) Untersuchung des Stereosehens im Zufallspunktmuster-VECP: Normbefunde und klinische Anwendung. Klin Monatsbl Augenheilkd 187, 205–208
von der Heydt R, Hänni P, Dürsteler M, Peterhans E (1981) Neuronal responses to stereoscopic stimuli in the alert monkey – a comparison between striate and prestriate cortex. Pflügers Arch 391, R34
von Helmholtz H (1853) Über einige Gesetze der Vertheilung elektrischer Ströme in körperlichen Leitern, mit Anwendung auf die thierelektrischen Versuche. Ann Phys Chemie 29: 211–233, 353–377
von Helmholtz H (1910) Handbuch der physiologischen Optik, Vol 3, 3rd edn. Voss, Hamburg
Wässle H, Peichl L, Boycott BB (1978) Topography of horizontal cells in the retina of the domestic cat. Proc R Soc Lond [Biol] 203, 269–291
Wernicke C (1874) Der aphasische Symtomenkomplex. Cohne und Weigert, Breslau
Wertheim T (1894) Über die indirekte Sehsschärfe. Z Psychol Physiol Sinnesorg 7, 172–187

Whitteridge D (1973) Projection of optic pathways to the visual corex. In: Jung R (ed) Handbook of sensory physiology, Vol VII/3B. Springer, Berlin Heidelberg New York, pp 247–268

Wilson FN, Bayley RH (1950) The electric field of an eccentric dipole in a homogenous spherical conducting medium. Circulation 1, 84–92

Woodworth RS (1938) Experimental psychology. Holt, New York

Woodworth RS, Schlossberg H (1955) Experimental psychology, 3rd edn. Holt, New York

Yanashima K (1982) Surface distribution of steady-state cortical potentials evoked by visual half-field stimulation. Gräfes Arch Ophthalmol 218, 118–123

Yarbus AL (1967) Eye movements and vision. Plenum, New York

Zihl J, von Cramon D (1986) Zerebrale Sehstörungen. Kohlhammer, Stuttgart

Zubek JP, Bross M (1972) Depression and later enhancement of the critical flicker frequency during prolonged monocular deprivation. Science 176, 1045–1047

Zubek JP, Bross M (1973) Changes in critical flicker frequency during and after fourteen days of monocular deprivation. Nature 241, 288–290

Central Processing of Sensory Information in the Control of Feeding

J. Le Magnen

Laboratoire de Neurophysiologie Sensorielle et Comportementale Collège de France, 11 Place Marcelin Berthelot, F-75231 Paris Cedex 05

1 Introduction

External sensory systems subserve behaviors. In this respect, sensory modalities fall into one of two categories. For example, vision and hearing, which are involved in a large variety of different behavioral responses, can be classified "generalists," whereas, its somesthetic systems (e.g., olfaction and taste) can be considered "specialists." Olfaction primarily subserves feeding behavior, while taste does so exclusively.

Progress in Sensory Physiology. Vol 8

This feeding specificity of the gustatory and olfactory functions is the product of natural selection, since feeding is essential to survival value and there is therefore a high selective pressure in favor of feeding. A point often neglected is that the differentiation of organs and functions in the evolutionary process occurs through adapted behaviors. Feeding and defensive behaviors are primarily involved in natural selection across and within species. Only individuals who are able to insure their feeding and self-defense survive to reach maturity and reproduce. By means of this selective pressure, the best sensory organs, brain processing of sensory input, and organs of motor output are selected and transmitted. Olfaction and taste are specifically adapted to subserve feeding behavior because they have been differentiated "through" feeding behavior.

Through evolution, the selective pressure of such behaviors results in a sort of a transgenerational memory. Anatomofunctional mechanisms are retained or eliminated from genetically transmitted programs. Likewise in individuals, beneficial sensorimotor links are fixed, and unserviceable ones eliminated in the course of associative learning. Neuronal organization or reorganization, known as "plasticity" of the central nervous system, is associated to these acquired sensory links. In each phenotype, behaviors and their contributions to body homeostatis result from genetically selected responses that are modulated by individual experiences.

2 General Organization of Sensory and Somatic Inputs to the Brain and of Their Control of Behavioral Output

2.1 Dual Organization of Sensory Systems

Modern neuroanatomical investigations have supported the notion (already presented by the author; see Le Magnen 1975) of a dual (neocortical and subcortical) organization of cephalic sensory systems. Subcortical visual pathways, such as the accessory optic tract, are distinct from the main geniculoneocortical visual system. They subserve specific neuroendocrine and sensorimotor functions (Marg 1964; Riley et al. 1981). A similar dual organization has been recognized in hearing with an extrathalamic inferior colliculus system primarily involved in the participation of acoustic signals in basic behavioral responses, such as spatial localization and echolocation (Erulkar 1972).

In the last decade, these dual cortical and subcortical systems have been clearly demonstrated in gustation by Norgren (1976, 1977, 1981). From the nucleus of the solitary tract (NTS), gustatory pathways relay in the parabranchial nuclei of the pons (PBN). From this relay, a contingent of fibers reaches the ventrobasal nuclei of the thalamus and, from this thalamic relay, the gustatory neocortical projection, which was recently well recognized and separated into two subunits (Yamamoto et al. 1980a, b). From the PBN, another contingent of fibers bypasses the thalamus and reaches the lateral hypothalamus and the substantia innominata directly. A more direct pathway connects the NTS to basolateral nuclei of amygdala (Saper et al. 1976; Ricardo and Koh 1978). It will be shown below that these subcortical and neocortical taste systems and their respective connections with motor programs of eating have distinct functions in feeding: the first is mainly involved in unlearned prewired sensory processing and responses; the second is mainly involved in the learning or relearning of these processings and responses. The same division is obvious in central olfactory projections; the predominant functional role is attributed mainly to the extrathalamic system. Direct projections from the prepiriform cortex to the lateral hypothalamus and from the amygdala to the preoptic area and the medial and posterior hypthalamus via the stria terminalis have been shown in neuroanatomical and electrophysiological studies (Scott and Leonard 1971; Scott and Pfaffmann 1972; Scott and Chafin 1975; Schwartz and Scott 1976). The olfactory thalamoneocortical projections seem to play a secondary role in the processing of the sensory input as judged by unchanged olfactory guided responses in decorticated animals (Beach 1940; Grill and Norgren 1978).

2.2 Centrifugal Pathways

Another general feature of the central organization of sensory systems is the existence and the functional role of centrifugal pathways. From the terminals of sensory projections, these pathways reach various sites of ascending fibers in a feedback system.

Neuroanatomic and electrophysiological studies have demonstrated visual centrifugal pathways in various species. Their terminals in the retina, which are well documented in low vertebrates, are unclear in mammals (Repérant et al. 1981). Centrifugal fibers (gamma fibers) to the olivocochlear bundle are well known (Keidel and Neff 1974).

In the taste system, centrifugal fibers from the paraventricular nuclei of the hypothalamus reach the dorsomotor nucleus of the vagus, which is closely connected both to the NTS and to the area postrema (Rogers et al. 1980). From the basolateral nuclei of amygdala, fibers reaching the NTS parallel afferent fibers (Ricardo and Koh 1978). Fibers descending from the NTS to tongue papillae have been hypothesized but have not yet been described neuroanatomically (Rapuzzi and Casella 1963; Sharma et al. 1977). The neocortical thalamocentrifugal pathways have been investigated (Yamamoto et al. 1980).

Centrifugal pathways reaching in the olfactory system the mitral-cell layer of the olfactory bulbs are well documented (Price and Powell 1970; Aguilar-Baturoni et al. 1976; Shafa and Meisami 1977). They mainly originate in the anterior olfactory nucleus, substancia innominata, amygdala and, through synaptic relays, the hypothalamus and the frontal cortex. Like in the gustatory system, we know of no olfactory centrifugal fibers that reach the peripheral organ. Therefore a modulation of olfactory input by centrifugal pathways cannot be ascribed to their action on the sensory transduction. Olfactory neocorticofugal fibers have not been studied.

There are some electrophysiological and neuroanatomical indications of links between gustatory and olfactory systems which seem to underlie their functional connections in feeding. Olfactory impulses have been recorded in single cells of the ventrobasal nuclei of the thalamus in which gustatory impulses have also been recorded (Keidel and Neff 1974). Single unit responses to odorants have been recorded in the NTS (Van Buskirk and Erickson 1977).

2.3 Reciprocal Interactions Between External Sensory and Somatic Inputs to the Brain in the Regulatory Control of Food Intake

Like other behaviors, feeding implies reciprocal interactions between the brain and the body, and between external and internal messages. Behavioral performances are "permitted," facilitated, or inhibited by neurally transmitted or blood-borne signals. It will be seen below that the systemic stimulus to eat or of hunger is now recognized as a blood-borne signal that is associated with glucoregulation and acts on glucosensitive target cells in the lateral hypothalamus. Coupled with this metabolic signal of hunger, parasympathetic and sympathetic afferent fibers from the stomach and intestine are involved in the satiation process brought about by the eating of food (Le Magnen 1983; Mei 1982). Once the hunger arousal is established by the internal blood-borne signal, a specific panel of external stimuli provided by foods is "promoted" in its capacity to quantitatively and qualitatively govern the eating motor program. In turn, the external sensory input and the behavioral performance modify the internal milieu through descending autonomic pathways and/or hormonal releases. We have demonstrated, for example, that oropharyngeal stimulations by food, induce a preabsorptive insulin release via the vagal innervation of the pancreas.

A combination of these reciprocal interactions between external and internal signals results in positive and negative feedback loops which, in feeding like in other behaviors, determine the behavioral pattern.

3 Feeding Mechanisms

3.1 Feeding Pattern

In the absence of a substantial store in the blood, the continuous tissue oxygen uptake, is directly dependent on a continuous intake through respiration. In contrast, the continuous tissue uptake of fuel and specific nutrients is only supplied intermittently by eating food. Feeding in all animal species is essentially periodical in nature.

A recording of free food intake of the rat (taken as an animal model) displays a dual periodicity (Le Magnen et al. 1980a; Le Magnen 1981, 1983, 1985) (Fig. 1). A rat that is offered its familiar balanced diet takes discrete meals of 1 – 6 g. These meals last 5 – 20 min and are separated by intervals of 50 – 200 min or longer; this is the "prandial periodicity." Large meals are separated by short intervals at night, smaller meals by longer intervals during the daytime. This is the superimposed dark-light, or circadian, periodicity.

This dual periodicity of the free food intake, as opposed to the fluctuating but continuous tissue uptake of fuel, has several implications with regard to providing body energy and maintaining nutritional balance. Continuous output from the blood requires an equivalent input into the blood. As feeding is an intermittent process, this continuous input is necessarily drawn from body stores; fat, glycogen, and the meal itself, which is considered a gastrointestinal store. This short-term supply of metabolites into the blood must satisfy another vital requirement: the relative constancy of the blood glucose content. Among the other

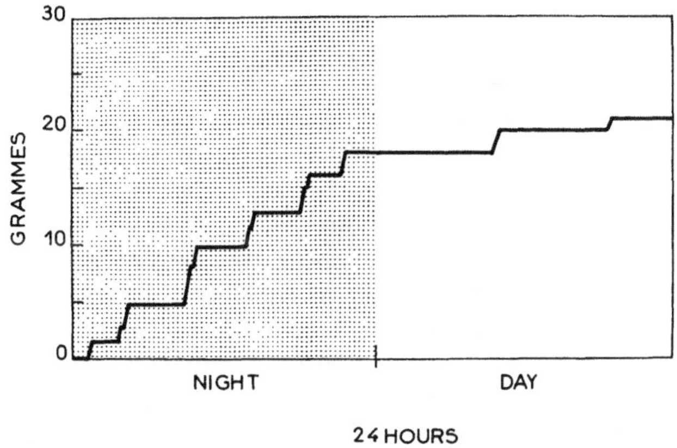

Fig. 1. Free feeding pattern of rats. A prandial and circadian periodicity is obvious

metabolites, e. g., fat, protein, and vitamins, only the constancy of blood glucose is maintained by endogenous mechanisms closely matching blood input to output. The question arises as to whether and how the onset of feeding and its own mechanisms are involved in this short-term regulation of blood glucose in addition to the endogenous regulatory mechanisms.

Another implication of feeding periodicity is its anticipatory nature. When a meal is set up and food is available in excess, is this onset triggered by a metabolic deficit? Once the meal has been engaged, is the amount of food eaten before satiation determined by the size of a previous deficit or does this meal intake anticipate further metabolic expenditures? An answer to these questions has been provided by a series of experiments in our laboratory.

In free-fed rats, a study of the relationships between the sizes of meals and meal-to-meal intervals in the two phases of the diurnal cycle yielded a major finding (Le Magnen and Tallon 1963, 1966; Le Magnen and Devos 1980). This analysis revealed that the size of meals during the dark and the light periods was not at all correlated to the time elapsed since the preceding meal or to the size of the last meal. On the contrary, there was a highly significant correlation at night between the size of meals and the duration of the postmeal interval that ended with the onset of a new meal. This postprandial correlation has been confirmed in various species (Balagura and Coscina 1968; Duncan et al. 1970; Kissileff 1970; De Castro 1975; Sanderson and Vanderweele 1975; Davies 1977; Wolf and Hainsworth 1977; Natelson and Bonbright 1978; Armstrong 1980; Marcilloux 1980; Ardisson et al. 1981), including man when he is deprived of external temporal cues (Bernstein 1976). Thus, two separate mechanisms seem to determine the time patterning and the cumulative intake. The first, which is responsible for the time of meal initiation, is dependent on previous nutritional events. The second, which is responsible for the amount eaten during a meal, is not. In ad libitum feeding conditions, the size of a meal is a store anticipating further metabolic expenditures.

During the night, relatively short intervals follow large meals, while long intervals follow smaller meals during the daytime. In other words, after the same amount of food has been eaten during a meal in the two periods, rats remain satiated at least twice as long during the daytime. This difference, which underlies the circadian periodicity of feeding, has been explained by the discovery of the diurnal lipogenesis-lipolysis cycle (Le Magnen and Devos 1970; Le Magnen et al. 1973; Le Magnen 1983). At night, a fat synthesis and therefore an increase of the body fat store are observed. They are correlated to the excess of the cumulative nocturnal intake over the concurrent metabolic rate, or energy expenditures. During the daytime, the intake, which is lower than the concurrent energy expenditures, is correlated to an observed fat mobilization and oxidation. At night, the utilization of food to simultaneously cover the current metabolic rate and fat synthesis accounts for the shortening of meal-to-meal intervals. During the daytime, the supply of an endogenous source of fuel by fat mobilization and a sparing of glucose utilization account for prolonged meal-to-meal intervals. In addition, a study of the relationship between the nocturnal positive energy balance and the daytime negative balance showed that rats, by an excessive intake and fat synthesis at night, anticipate the deficient intake and the mobilization during daytime (Le Magnen and Devos 1982, 1984) Likewise, hibernators anticipate hypophagia and hibernation with hyperphagia and fat synthesis. Migrating birds do the same before migration (Odum 1960; Mrosowsky and Sherry 1980).

In contrast to this feeding pattern observed when food is available in excess, the size of the first meal following food deprivation and the rate of initiation of subsequent meals are dependent on the duration of deprivation (Le Magnen 1977b; Le Magnen et al. 1980a). However this deprivation-induced feeding differs according to the time of day, i.e., it is higher at night than during daytime.

3.2 Mechanism of Meal Initiation: Systemic Stimulus of Hunger Arousal

Experimental studies in the rat have confirmed an earlier suggestion that hypoglycemia, as reflecting a fall of tissue glucose availability, is critically involved as a systemic stimulus to eat.

A microflow of blood was drawn for several hours via an intrajugular catheter in a free-moving rat. Blood-glucose level and free-food intake were continuously and simultaneously recorded. It was shown that the blood glucose was perfectly constant from meal to meal. However, the start of every meal was preceded by a slight fall in blood-glucose concentration. This hypoglycemia occurred 3 – 5 min before the meal onset and continued until the postabsorptive rise and the subsequent return to the regulated level (Campfield et al. 1985; Louis-Sylvestre and Le Magnen 1980) (Fig. 2). An injection of either glucose (Campfield and Smith 1985), glucagon (Geary et al. 1981; Balagura et al. 1975), or adrenalin (Lagaillarde and Louis-Sylvestre 1985) at the predicted time of onset of a meal retarded the meal by antagonizing the fall in blood glucose. After glucose or adrenalin injection, the retarded meal was begun when blood glucose once more reached a level lower than the meal-to-meal constant level (Campfield and Smith 1985; Lagaillarde and Louis-Sylvestre 1985).

Further evidence of the involvement of hypoglycemia in meal initiation was drawn from the time dependency of deprivation-induced hypoglycemia throughout the day (Larue-Achagiotis and Le Magnen 1985b). The fall in blood-glucose concentration induced by 3 h food deprivation was measured in eight groups of rats. It was shown that the hypoglycemia induced by the 3 h deprivation throughout the

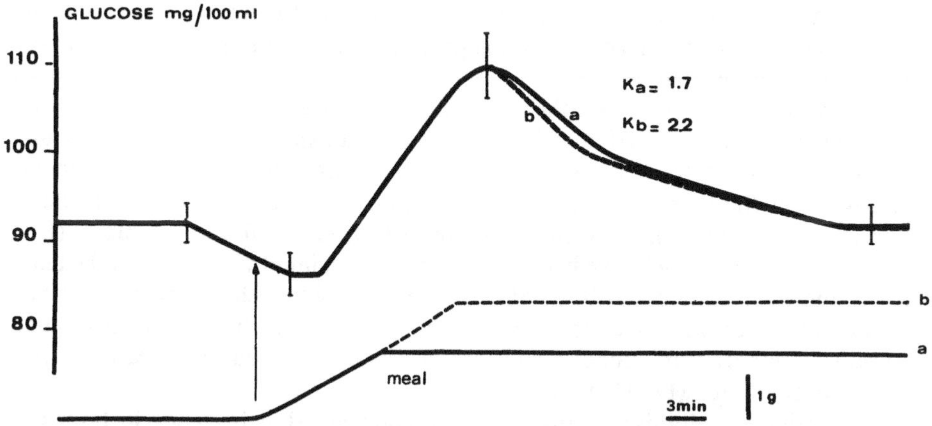

Fig. 2. Preprandial and prandial evolution of blood-glucose level during night (*a*) and daytime (*b*). *K*, glucose utilization coefficient during the fall of the prandial hyperglycemia. Glucose tolerance is higher during the night than during daytime

24 h was correlated to the 3 h intake when the same rats were fed ad libitum. For example, a 3 h food deprivation at the start of the dark period induces a steep fall in blood glucose, and rats eat a large amount of food during these hours. The removal of food during the first 3 h of the light period, however, does not induce a fall in blood glucose, and the rat does not eat at all at this time even if food is available. From this result, one could argue that the rat eats throughout the day at a rate just required to maintain its blood glucose level and that a meal is started by a fall in blood glucose.

This premeal fall in blood glucose in free-fed rats is not graded as a function of the time elapsed since the preceding meal and is the same whatever the size of the started meal. This confirms that the amount eaten during a meal is not determined by the level of a metabolic deficit present at the time of its initiation. The premeal hypoglycemia is only a triggering signal. The eating of a meal is the filling of a store. This store, through the intestinal absorption, feeds the liver, which regulates the input into the blood. When the gastrointestinal store reaches a given low level, the time of which varies as a function of the rate of the metabolic food utilization (rapidly at night, slowly in the daytime), the condition of meal onset is reattained.

The effect of a continuous insulin infusion confirms this mechanism (Larue-Achagiotis and Le Magnen 1985a). Insulin, by augmenting the tissue glucose uptake (and particularly the tissue adipose uptake), augments the rate of food utilization. In addition, an insulin infusion during the daytime abolishes lipolysis, which normally occurs at this time, and induces fat synthesis. As might be expected, this continuous intravenous insulin infusion induces hyperphagia by augmenting the frequency of meal initiation mainly during daytime.

After 3 h of food deprivation in the above experiment, the size of the first meal and the subsequent intake when access to food was restored were shown to be proportional to the variable hypoglycemia induced by the deprivation throughout the day (Larue-Achagiotis and Le Magnen 1985b). Thus, the strength of the systemic stimulation to eat, which affects the size of the meal after food deprivation (in contrast to the free-feeding condition), is proportional to the level of the deprivation-induced hypoglycemia.

This series of data, as well as others (which cannot be developed here), lead to the conclusion that three different interacting mechanisms determine the amount and time pattern of food intake. The first mechanism is that of the systemic stimulus to eat (hunger) in which the signal appears to be the decrease in tissue glucose disposal. None of the other variations of other metabolites in blood (fats and proteins, for example) have been shown to be associated with food intake temporarily and quantitatively. A body of evidence provided by electrical stimulation, self-stimulation lesions, or electrophysiological recordings strongly suggests that the brain targets of this hunger signal are glucosensitive neurons located in the lateral hypothalamus (Le Magnen 1983).

An independent mechanism is involved in regulating, that is to say in limiting, both the excessive depletion and repletion of the body fat energy store. This mechanism indirectly affects the feeding mechanism proper by increasing the rate of meal initiation during fat synthesis and decreasing it during fat mobilization. it is responsible for the circadian periodicity of food intake. The ventromedial nuclei of the hypothalamus govern this regulation of body fats and body weight. It is well known that their destruction leads to sustained hyperphagia, to the disappearance of the circadian periodicity of food intake, and to obesity (Le Magnen 1983, 1984).

The third mechanism or control system is that which determines the amount of food eaten from the start to the spontaneous end of a meal and the selection of foods. Unlearned or learned responses to the sensory activity of foods – mainly to their taste and smell – are decisive factors in this mechanism.

3.3 Palatability of Foods as Determinants of Meal Size and of Food Selection

Mediation of food intake by oropharyngeal sensory activities of foods becomes evident when various food items are substituted for the usual and familiar food. In the ad libitum condition or in a given state of food deprivation, a meal is initiated or not and, if initiated, in rats varies from 1 to 6 g or more, depending on the food offered. This responsiveness to the sensory properties of food is termed the palatability of the food.

The palatability of food is sensory specific, that is to say, specific to the particular sensory activity (taste and smell) of each food item. This was demonstrated earlier in the following experiment (Le Magnen 1956): Rats were habituated for 32 days to a 2 h morning meal. This balanced synthetic diet was flavored by adding a trace of one of four different odorants: A, B, C or D. After some days, the 2 h intake was identical whichever odor labeled the food. When the four flavored forms of the same diet were presented successively during the morning meal, 30 min each, the meal was considerably augmented. In a similar experiment, three or four flavored versions presented simultaneously instead of successively also stimulated overeating (Le Magnen 1960). This specificity of the sensory component of stimulation to eat explains the fact that rats offered ad libitum a choice of very palatable foods (cafeteria regimen) became hyperphagic and obese (Sclafani and Springer 1976; Rolls et al. 1980; Louis-Sylvestre 1984). It has been shown that this hyperphagia is due to over eating in which rats successively eat the palatable foods (Rogers and Blundell 1980; Le Magnen and Devos unpublished observations). In the single-food presentation or in a choice, sensory specificity is also demonstrated by the fact that changing the food by adding noncaloric additives (artificial sweetener, bitter compound) enhances or reduces the response compared with a reference item.

When compared to a control, these enhanced or reduced responses are commonly termed preferences and aversions. However, this terminology is often misleading. In fact, palatability is a continuum, and there is no reason to distinguish between positive and negative palatability. A true aversion, which is a defensive rejection of food, must be distinguished from unpalatability or relative aversion.

The palatability of foods and their sensory specificity are "state dependent"; in other words, the response to sensory stimuli from foods – and only to those stimuli – is promoted by the hunger state. Rats offered, after an insulin injection, a choice of water, isotonic NaCl, and saccharin solutions enhanced their 30 min intake only of the sweetened solution (Le Magnen 1953). Similarly, rats that have been deprived of foods augment their intake of a saccharin solution and not of water or of a salty solution. When thirsty or sodium-depleted (after adrenalectomy, for example), they selectively enhance their intake of water or of the salty solution even when the latter is made of a nonsodium salt (Shulkin 1982; Berridge et al. 1984). Thus, the sensory-specific response to the food or to the fluid is not only state dependent; it is also state specific.

In man, the subjective correlate of food palatability is the "pleasantness" of food stimuli. It has been known since Aristotle that this pleasantness of foods increases with hunger and is suppressed by the abolition of hunger, or satiation, at the end of a meal. An objective technique has recently been developed to measure the palatability of foods in man (Pierson and Le Magnen 1969; Bellisle 1979). The recording of the chewing and swallowing pattern during a spontaneous meal showed that some features of the pattern, such as duration and number of masticatory movements per mouthful, varied with the palatability of foods. These indices were correlated with the subjective judgment of pleasantness and with the particular size of the meal and varied with each food item. When measured in this way, it was shown that the palatability decreased from the beginning to the end of the meal and that the initial level was augmented as a function of a previous food deprivation (Bellisle 1979; 1984; Bellisle et al. 1984) (Fig. 3).

Thus, the cumulative intake of food from meal to meal is dependent on the additive or synergistic combination of an internal hunger signal and of the specific response to each food item to its sensory activity. This combination is probably based upon the convergence in the lateral hypothalamus of olfactory and gustatory projections to sites responding to the blood-borne signal of hunger. It will be seen below that discharges in the monkey and rodents of single neurons in this area are changed by the sight or the taste of a food only when the animal is hungry (Rolls et al. 1980).

The last point to be clarified is the exact influence of the initial facilitatory action of the oropharyngeal stimulation in determining the amount eaten during a meal before the onset of satiety. The satiation process is brought about by a progressive

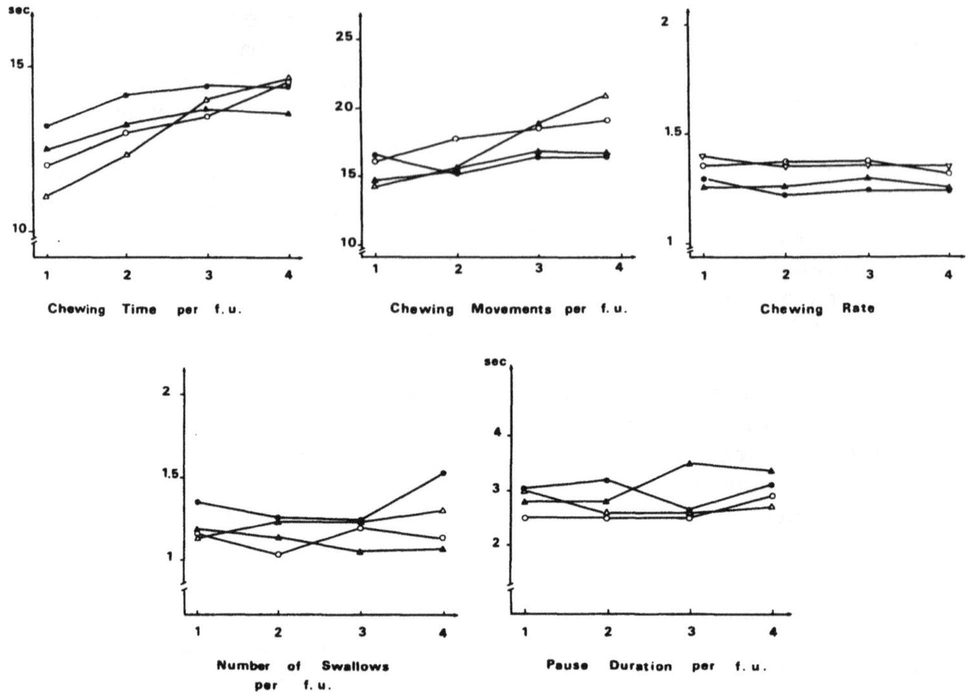

Fig. 3. Temporal evolution of 5 parameters of the chewing-swallowing pattern during the four successive quarters of the meal (1-2-3-4); some of them are correlated to the palatability of the food as assessed by meal size and subjective scaling, and indicate a progressive lowering of food palatability from the beginning to the end of the meal. △, high palatability, long starvation; ○, high palatability, short starvation; ▲, low palatability, long starvation; ●, low palatability, short starvation

inhibitory or hunger-suppressing activity of the food accumulating in the stomach and/or in the intestine (Le Magnen 1976, 1983). The balance between the respective strength of the oral positive and post-oral negative feedback determines the amount that has been eaten at the termination of the meal. Since the palatability or sensory stimulation to eat is hunger dependent, the progressive shift from palatability to unpalatability during the meal may be explained by the postoral negative feedback. However, there is different evidence to show that the cumulative effect of the sensory activity at the oral level contributes to the negative feedback. In addition to a sensory-specific stimulatory effect, each food seems to possess a sensory-specific satiating power (Booth 1972). Overeating by humans and in the above-mentioned experiment in rats may be interpreted as demonstrating both the sensory-specific stimulation and the sensory-specific satiation.

Finally, a major question is raised. In a laboratory animal, such as the rat, and in man as well, a meal is terminated before a substantial intestinal absorption of the nutritive content of the food occurs. Thus, in order to be adjusted both to the nutritive properties of food and to nutritional demand, the oral intake must anticipate these properties and demands on the basis of the sensory activities of the food. This adjustment does in fact exist, i.e., is performed. The amount and choice of various foods eaten are roughly adjusted to the nutritive activity of the food and to metabolic requirements without delay or after some familiarization. How is this possible?

Various conditions must be satisfied for nutritive properties to be predicted on the basis of oropharyngeal cues. A first condition lies in the nature of the nutritionally and sensory active compounds in the diet. The carbohydrate content of the diet may be sweet. A toxic compound present in the diet may be responsible for bitterness; but specific taste or smell of the three macronutrients, vitamins, oligoelements, and so on does not exist. Olfactory properties of most food items, fats, and proteins are due to traces of odorous compounds and not to the metabolically active content of the diet. These odorous natural additives have no quantitative or qualitative relation with the nutritive properties of the diet to which they are associated.

When the metabolically active compound is also the sensory stimulus, another condition for an anticipatory response to the food is that the qualitative coding by the sensory apparatus involved corresponds to a biochemical coding. This condition would be achieved if carbohydrates – and only they – were sweet and if their sweetness predicted their caloric density, of all toxic compounds – and only they – were bitter, and so on. However, as is well known, though most sugars are sweet, other carbohydrates are not and nonmetabolizable substances are also sweet. Metabolically active NaCl cannot be discriminated from toxic LiCl. Many toxic compounds are not bitter, and many bitter compounds are not toxic.

In the following paragraphs, it will be pointed out that unlearned anticipatory responses of palatability are manifested when and inasmuch as there is a correspondance between the qualitative sensory coding and the nutritive properties. When there is no such correspondance, it will be seen that the genetically programmed response is modified or a new response is acquired by learning.

4 Taste Discrimination and Taste Response

The qualitative coding of various gustatory stimuli in four different classes is partially a genetically programmed biochemical coding. Natural sugars (carbohydrates and suppliers of calories) are sweet. As just mentioned, inasmuch as the taste quality corresponds to a biochemical entity, there are genetically adapted, ingestive responses.

4.1 Central Coding of Taste Information

Studies of the qualitative and quantitative coding in the taste apparatus have confirmed the discrimination of four basic taste qualities by identifying the neural mechanism of this discrimination. Single fiber analysis in chorda tympani and glossopharyngeal nerves, and single recordings in the NTS and PBN showed responses of most fibers and cells to more than one class of stimuli (Pfaffmann 1974), but these fibers and cells could be classified as responding maximally or best to one of them (Frank 1977; Frank and Pfaffmann 1969). In addition, a chemotopic organization of best responses to sweet, acidic, salty, and bitter stimuli on the tongue as well as in the NTS and PBN has been recognized (Halpern 1965; Scott 1974, 1977,1984; Scott and Perrotto 1980).

Confirming the early suggestion by Pfaffmann, it has now been clearly demonstrated that the qualitative discrimination results from across-fiber patterns of discharges which differ for the four basic stimuli. These across-fiber patterns are not identical but similar for stimuli whose tastes are judged identical or close to one of the four basic standards (Scott 1974, 1977; Frank 1977; Smith et al. 1979; Di Lorenzo and Schwartzbaum 1980, 1982; Scott and Perrotto 1980). The analysis of the recorded across-fiber patterns at different levels of gustatory input has made it possible to quantify these differences and similarities. Correlation between all pairs of recorded neurons and of their responses to a series of stimuli established that the across-fiber patterns of sweet compounds (including saccharin) are highly positively correlated. They were all negatively correlated to

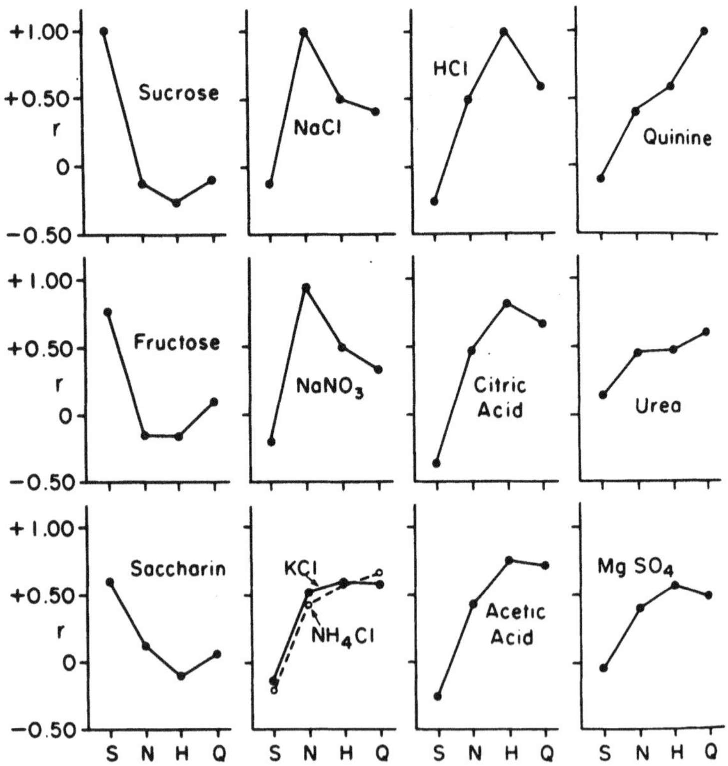

Fig. 4. Correlation taste profiles across four tastes: 0.1 M sucrose (*S*), 0.03 M NaCl (*N*), 0.003 M HCl (*H*), and 0.001 M quinine hydrochloride (*Q*) for the responses (number of nerve impulses in 5 sec.) of 46–79 hamster chorda tympani nerve fibers to these tastes and 0.3 M fructose, 0.001 M Na-saccharin, 0.03 M $NaNO_3$, 0.03 M NH_4Cl, 0.1 M KCl, 0.003 M acetic acid, 0.001 M citric acid, 1.0 M urea, and 0.03 M $MgSO_4$. The *r*'s plotted are Spearman rank-difference correlation coefficients. The coefficients for responses to sucrose and sucrose and for each taste and itself were assumed to equal +1.00. From Frank (1977) with permission of the Rockefeller University Press

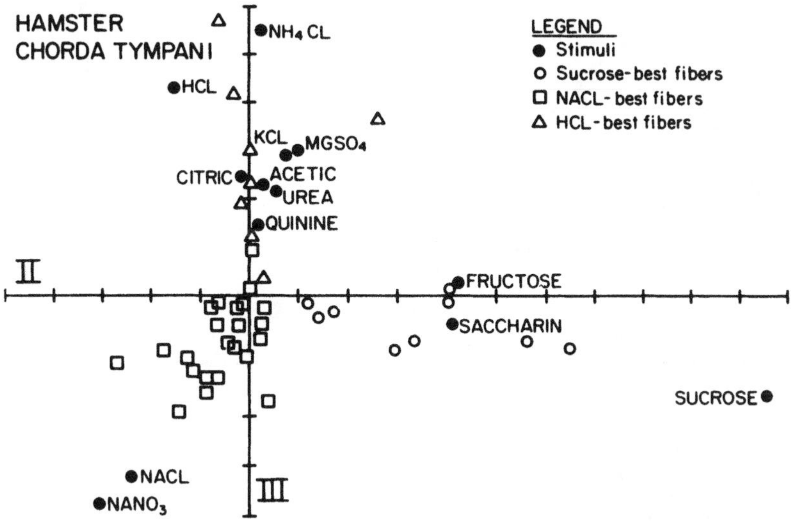

Fig. 5. The Schiffman-Falkenberg taste space developed from the responses of 40 hamster chorda tympani nerve fibers to 0.1 M sucrose, 0.03 *M* NaCl, 0.003 *M* HCl, 0.001 *M* quinine hydrochloride, 0.3 M fructose, 0.001 *M* Na-saccharin, 0.03 *M* $NaNO_3$, 0.03 *M* NH_4Cl, 0.01 *M* KCl, 0.003 *M* acetic acid, 0.001 *M* citric acid, 1.0 *M* urea, and 0.03 *M* $MgSO_4$. Both the stimuli (represented by *solid circles*) and the fibers (represented by *outline symbols: circles,* sucrose-best; *squares,* NaCl-best; and *triangles,* acid-best) are plotted in the same Euclidean space. Factor II is plotted along the abscissa, factor III along the ordinate, but factor I, which only accounts for differences in the size of responses across stimuli and fibers, is not plotted. From Frank (1977)

across-fiber patterns generated in the same units by acid, salty, and bitter compounds, while the latter three classes of compounds were also dissociated (Fig. 4). A multidimensional analysis of these patterns showed that clusters corresponding to the four classes were subjectively and behaviorally discriminated (Fig. 5). An identical across-neuronal pattern is generated by a particular stimulus in the chorda tympani, the NTS, and the PBN.

The intensity coding is represented by the total discharge of the series of neurons involved in the qualitative coding and may be quantified by using the area-under-the-curve envelop of the fiber pattern (Erickson 1963; Erickson et al. 1965). As verified by Ganchrow and Erickson (1970), the profile of this envelop, that is to say the basis of the qualitative coding, is unchanged at various concentrations of the stimulus (Ganchrow and Erickson 1970; Doetsch and Erickson 1970) (Fig. 6). Interestingly enough, saccharin is an exception.

Augmenting its concentration produces a shift of the across-fiber pattern from the sweet to the bitter-like pattern. This corresponds to the change in the perceived taste of saccharin and of the behavioral response of animal species with increasing concentrations (Carper and Polliard 1953). The total discharge increases as a function of concentrations according to psychophysical determinations of the relationship between perceived intensity and stimulus concentrations. The exponent of the law is different for the two classes of stimuli.

In this clear picture of outstanding data, some points remain questionable and questioned. Two opposing theories have been put forward in the past steps of the

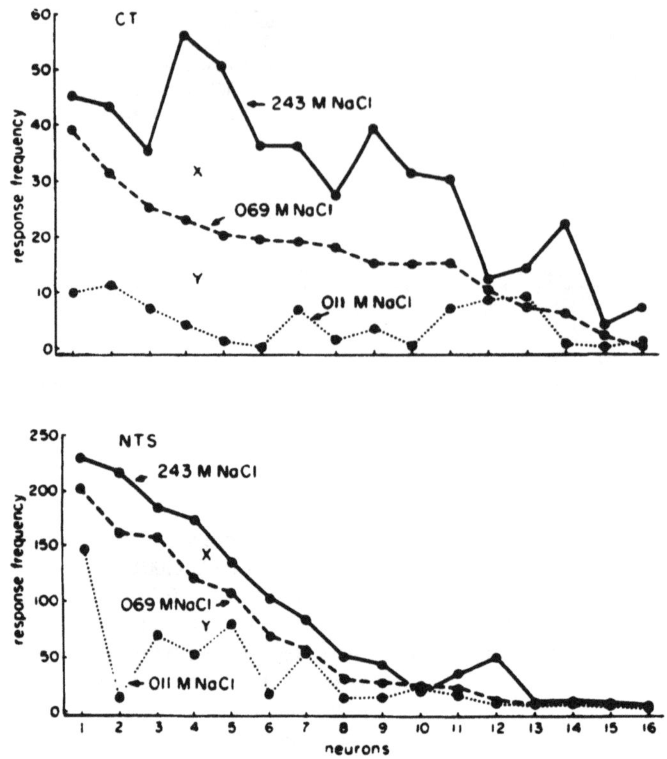

Fig. 6. Across-fiber patterns during the 2nd sec at the CT and NTS levels to three concentrations for NaCl. Each data point represents the frequency of response for the labeled neuron calculated from one or two stimulus presentations. The strongest and the weakest concentrations were each six jnd's for humans from the middle concentration. *X* represents the area between the stronger two concentration curves; and *Y*, the area between the weaker two. The form of the across-fiber pattern changes moderately (see Fig. 4) across concentrations. The corresponding number labels for CT and NTS neurons are not intended to imply anatomical or functional relations. (Ganchrow and Erikson 1970)

above investigations. One theory supported the idea that the qualitative coding was due only to the existence of best responding fibers or labeled channels of information; it was opposed to the theory of the across-fiber pattern. Authors seem now to agree that the two theories complement each other. Another point is the exact role of the chemotopic organization in afferent nerves and in the two central synaptic relays. It seems that this spatial organization of fibers and of cells best responding to "bitter" stimuli and so on is a complementary and necessary aspect of the across-fiber pattern for qualitative discrimination. Computation of the correlation coefficient or the mathematical multidimensional analysis of cell responses is not, of course, performed as such by the brain. Brain analysis of the gustatory afferent pattern is likely to be tridimensional, that is to say a spatial analysis based upon the chemotopic organization. Some data suggest that a temporal factor (rate of initial discharges) which is similar within, and different between, the four classes of stimuli could also be involved.

This addition of a spatiotemporal discriminative performance to labeled channels and to an across-fiber pattern seems necessary for an understanding of the responses to mixtures of stimuli. In mixtures of sweet and bitter stimuli, sweet and acidic stimuli, and so on, various combinations of intensities are subjectively observed. However, the two qualities remain discriminable and can be recognized. Evidence has been provided that these effects of mixtures are not due to the competition between molecules at the level of peripheral receptor sites, but to an interaction of neural messages generated by the two mixed stimuli (Lawless 1979). When rats are made Na deficient by adrenalectomy, they augment (in contrast to controls) their intake of a mixed NaCl-sucrose solution. The intensity compensation is manifested by the fact that a fivefold increase of the concentration of NaCl in the mixture is needed to induce the same volume intake as a pure NaCl solution in the same adrenalectomized rats (McCutcheon and Brown 1983). A random distribution of cell responsiveness in the NTS cannot explain this observation. Suppose we have three neurons, two of which are highly responsive to bitterness and weakly to sweetness. A mixture of the two stimuli should give the same firing rate in the three neurons and therefore should make the two stimuli in discriminable in a mixture. On the contrary, the localization in different parts of the NTS of neurons best responding to sweetness and of those responding to bitterness can explain both the intensity combination and the maintained qualitative discrimination in the mixture.

In the thalamus and cortical projections, the qualitative coding and its relation with the biological significance of stimuli is more ambiguous than in the upstream steps of the central processing.

The hunger dependency and specificity of palatability responses based upon taste stimuli are correlated to the same dependency and specificity of taste responses recorded in the central gustatory pathways. Multiunit discharges recorded in the NTS to sucrose, NaCl, and quinine are selectively attenuated in food-deprived rats by the inflation of a balloon in the stomach. Responses to sucrose are maximally suppressed; responses to quinine are not affected (Glenn and Erickson 1976). The sate dependency of the response is demonstrated at terminal sites of the diencephalic projection of the gustatory input. In a particular and significant study, single cell responses were recorded in the hypothalamus under stimulation of the tongue either by water, sucrose, or quinine (Norgren 1970). Visual and sound stimulations were also performed. About half of investigated cells responded to at least one of the gustatory stimulations. Animals stimulated themselves through electrodes placed in sites where sucrose responses were recorded, and not in sites where quinine responses were recorded. The former are situated mainly in the medial forebrain bundle and separated from the others found in other regions of the hypothalamus. In units responding to sucrose, the probability of obtaining the response was augmented under food deprivation while the same units responded only to water in satiated animals.

Similar responses to the sight and the taste of foods as a function of hunger and satiety and of the biological significance of stimuli have been studied in the monkey (Burton et al. 1975, 1977; Rolls et al. 1976; Mora et al. 1976). In the lateral hypothalamus (LH) and substantia innominata, the resting firing rate of single units is increased or decreased when the monkey is hungry and is tasting or looking at its food. The firing rate returns progressively to the resting level during the meal.

The state dependency of such central responses to gustatory stimulations and the associated behavioral responses have often been interpreted theoretically as a result of a direct action of blood content (blood glucose or Na levels) on taste cell receptors or on the NTS. With regard to the tongue receptor cells, it is clear that their responsiveness to sweet substances is not modified by an insulin-induced hypoglycemia, and their responsiveness to salts is not modified by adrenalectomy as assessed by recordings in the chorda tympani (Pfaffmann and Bare 1950; Pfaffmann and Hagström 1955). A reduced response to sucrose in the NTS and an equivalent reduction of a behavioral response to sucrose intensity were recently demonstrated after hyperglycemia induced by IV. Glucose infusion (Gisa and Scott 1983, 1986). However, this effect is likely due to centrifugal pathways coming from hypothalamus sites responsive to the blood glucose level rather than to a direct action on the NTS processing. Electrical stimulations of LH reduce the NTS responses to taste stimuli supplied to the tongue (Bereiter et al. 1980; Matsuo et al. 1984). The same LH stimulation activates basolateral nuclei of amygdala connected to the NTS and involved in palatability responses (Rolls 1972).

4.2 Unlearned and Learned Feeding Responses to Taste Stimuli

Feeding responses to taste which differ between the four classes of stimuli are based upon this neurofunctional organization of the gustatory system. As mentioned above, inasmuch as taste discriminations correspond to distinct biochemical properties of stimulating compounds, genetically programmed distinct responses are manifested which are both state and taste specific. Sweet substances are palatable and spontaneously preferred to water as early as the first day of life in animal and human newborns (Desor et al. 1973; Steiner 1973; Pfaffmann 1974). This responsiveness to sweet solution is augmented by hunger (Le Magnen 1953; Hsiao and Tuntland 1971; Strouthes 1971). A preference-aversion function to NaCl solutions of various concentrations is exhibited by rats with a maximal preference for an isotonic solution (Young and Falk 1956). This maximal palatability of the isotonic concentration and the amount drunk are changed both by altering the blood osmotic pressure and by Na deficiency (Richter 1936; Le Magnen 1977a; Shulkin 1982). Acid and bitter compounds are universally aversive; they induce rejection. However, these innate responses are inadequate when sweet substances are not sugars, salty compounds not sodium salts, etc. The question may be raised as to the origin of this lack of response or defect in the correspondence of the sensory qualitative coding and biochemical properties of stimuli. It could be due to the fact that artificial sweeteners do not exist in nature. If the synthesis of saccharin had not been invented by Remsen and Fahlberg and were also performed in some nonedible or toxic plants, saccharin would presumably habe been tasteless or bitter. If LiCl had been as available in nature as NaCl, the danger of poisoning would have made it necessary that a mechanism discriminating between the two compounds be differentiated by natural selection in the evolutionary process.

Upon the background of unlearned responses, differential responses are acquired in a conditioning in which nutritive and nonnutritive properties of the compound act as a primary reinforcer or unconditioned stimulus (UCS).

The palatability of saccharin solutions, for instance, is either reinforced or extinguished by this conditioning. In a significant experiment, two groups of rats were offered a daily short presentation of a saccharin solution. In one group the intake was followed by a gastric load of a sucrose solution, in the other by a water load. After 14 days, despite the inversion of gastric loads in the testing day, the intake of saccharin was considerably higher in the first group than in the second, in which the initial preference for the saccharin solution had been extinguished (Capretta 1964). In a similar experiment, two groups of fasted rats were presented a short daily meal containing either a sucrose or a saccharin solution. From day to day, the intake of the sucrose group increased. In the saccharin group, after an initial increase, the intake rapidly fell to zero on the first day of fasting (Le Magnen 1954).

The reduced intake of a diet following its adulteration by quinine is maintained for a long time because quinine is toxic (Kratz et al. 1978a, b; Aravich and Sclafani 1980; Hayback and Boyle 1982). The same initial reduction in intake due to the addition of sucrose octoacetate to the diet disappears rapidly because this bitter compound is not toxic. Rats presented with a choice of sucrose or fructose solutions initially preferred the sweeter one. They rapidly shifted their preference to the calorically denser solution (Jacobs 1962). More dramatically, the high palatability value of a sweet solution, be it caloric or not, and the preference for the most preferred NaCl solution can be reversed by the well-known illness-conditioned aversion procedure. Pairing the intake of a saccharin solution with a postingestive injection of LiCl induces a strong and persistent aversion to the saccharin which is generalized to some other sweet solutions (Rozin and Kalat 1971; Nowlis et al. 1980). The oral intake of a LiCl solution is followed by an aversion which is generalized to a NaCl solution (Nachman 1963).

We shall come back below to the brain mechanisms involved in such conditioning of palatability responses and of aversions through postingestive activities of the orally ingested item. The question arises as to whether this acquired discriminative response is accompanied by a learned new central processing of the afferent sensory input. Of course, such discriminative responses between identically tasting solutions can be acquired on the basis of other sensory cues, for example a fixed position of each of the two items. However, some data suggest that some rearrangements of the central qualitative coding by learning can occur and underlies the acquisition of discriminative responses. The peripheral message offers the possibility of a learned alteration of the genetically programmed discriminative process. Evidence has been provided by cross-adaptation and also in electrophysiological and configurational studies that the sensitivity of taste cells to various sweet substances is due to multiple receptor sites (Dugas du Villard et al. 1981; Schiffman et al. 1981; Faurion 1982).

As an effect of the different pattern of action of various sweet substances on these multiple receptor sites, the across-fiber pattern they generate, though similar, is not identical. It is possible that a contrast between these similar patterns is accentuated through learning at the level of the NTS or higher in the ascending pathways. There is no evidence that saccharin, made aversive by illness-induced aversion, becomes bitter or even is no longer perceived as sweet; but recent results suggest that a rearrangement of single cell responses in the NTS indeed occurs after such a conditioning (Chang and Scott 1984). This would facilitate the specificity of the learned aversion. However, the learned aversion to saccharin, as mentioned above, is generalized to some other sweet compounds. The degree of this generalization indicates the limit of this learned rearrangement of the qualitative coding (Nowlis et al. 1980).

5 Olfactory Discrimination and Responses to Foods

5.1 Discrimination of Food Odors

Discrimination of the smell of foods does not provide – not even roughly – discrimination or recognition of their nutritive properties as may be the case with taste. There are various reasons for this lack of a genetically programmed olfactory analysis of the environment corresponding to an analysis of the nutritive properties of stimuli. In addition to food seeking and food intake, the olfactory system, in contrast to taste, subserves other behaviors, such as sexual and defensive behaviors. Nevertheless, a qualitative coding could have been differentiated for the detection of some vegetal or animal food odors through specialist receptors, as is the case in insect species for the detection of pheromones. Such food specialist olfactory receptors are unknown, at least in omnivorous higher vertebrates. However, peripheral olfactory receptors responding only to food plant odor have been recognized in insects. Likewise, in the same or other species taste receptors have been found which respond only to one particular sugar, this receptivity corresponding both to the species intake, carbohydrate metabolism, and the biotope of the species (Le Magnen 1977b).

Electrophysiological investigations and, more recently, the technique of 2-DG labeling have demonstrated that, at peripheral and bulbar levels, the molecular analysis is performed on the basis of across-fiber patterns specific to each discriminated molecule (Holley and MacLéod 1977; Shepherd 1977; Lancet et al. 1982). In contrast to taste, similarities of these single molecule patterns and the correlation of these similarities to common biochemical properties of stimuli have not been recognized. Again in contrast to taste, this individual recognition of odorous molecules generally disappears in mixtures. An unpredicted odor of the mixture which is different from that of the components is generated by the combination of the various across-fiber patterns.

These two characteristics of olfactory discrimination of pure molecules and of mixtures create the particular conditions for the discrimination of food odors. The specific odor of each food item is generally due to a mixture of traces of a tremendous number of odorants which are permanently or occasionally associated with food products. Each fruit, each sort of meat or fat, etc., can be individually identified. This identification cannot be a recognition of the common nutritive properties of a series of products. For, in contrast to the taste of sugars, NaCl, and so on, the odor labeling of food is not due to its nutritive component but – as mentioned earlier – to the associated natural odorous additives. Most pure carbohydrates are odorless and there is no specific fat or protein odor. In humans, cooking and "cuisine" add to this infinite variety and complexity of food odors and to the lack – before experience – of relations between perceived odors and postingestive activities of the food.

These conditions for olfactory discrimination preclude the existence of genetically programmed responses to food which are adjusted to nutritive properties and metabolic demands. Such responses have been described in low vertebrates (snakes, for example) but have not been observed clearly in newborn mammals (Burghardt 1970). Nevertheless, it has been reported that olfactory guided behavior during the sucking period is involved in the nipple attachment of newborn rats (Blass et al. 1977; Teicher and Blass 1977; Teicher et al. 1978). In humans, the recording of facial expressions in response to odor stimulations in newborns seems to demonstrate innate reactions of preferences and rejections. Yet a clear relation of these responses to the food-related significance of stimuli is not obvious (Steiner 1973).

However, there have been ample demonstrations of how the palatability of foods is based upon olfactory cues and adjusted to nutritive properties and nutritional status. They are learned responses.

This odor conditioning of palatability was shown long ago by Harris et al. (1933), who used the technique of odorized food. Rats were made vitamin B_1 deficient with a vitamin-free diet. They were then offered a choice between two forms of the same diet: a vitamin-free diet or a diet enriched with a minute amount of the odorless vitamin and labeled with an arbitrarily chosen odorant. After some days, rats ate exclusively the vitamin-rich and odorized diet. When the odorous marker was shifted to the vitamin-free diet, rats followed the odor and sometimes chose to eat exclusively the rachitogenic diet.

Later, two typical experiments demonstrated how the odor-guided appetite was adjusted to, and conditioned by, the caloric properties of the diet. Rats were offered two 60 min daily meals. The same synthetic diet was offered alternately in these two meals labeled by adding odorant A or B. After the oral intake of the form A in one group and of the form B in another, rats were injected i. p. with a glucose solution adjusted after each meal to 25% of the calories just eaten during the meal. The oral intake of the other form of the diet was followed by a NaCl control injection. After some days, a higher intake of the form without glucose supplement developed and was manifested in a final choice between the two odorized forms of the diet (Le Magnen 1959) (Fig. 7). In a similar experiment, rats were presented alternately with two solutions of starch made discriminable by the addition of two odorants. One solution contained 5%, the other 60% starch. After some days, a solution of an intermediary concentration (35% starch) was separately labeled with either of the two odorants. Rats then drank a larger amount of the solution labeled with the odor that had previously been associated with the calorically less dense solution (Booth 1972). In the two experiments, rats preferred the nonsupplemented diet or the low caloric diet solution. Likewise, they rapidly augmented their intake of the familiar diet which was calorically diluted by inert material. Comparable results have been reported in human subjects (Booth et al. 1982).

The question may be raised as to whether this conditioning by caloric density is that of a differential initial facilitation or of the combined orogastrointestinal satiating process. A conditioning of the initial facilitation is supported by many other experiments. Such a conditioning occurs even in utero (Smotherman 1982; Stickrod et al. 1982) Injecting an odorant in the amniotic fluid induces a preference for a food labeled with this odorant during the postweanling period.

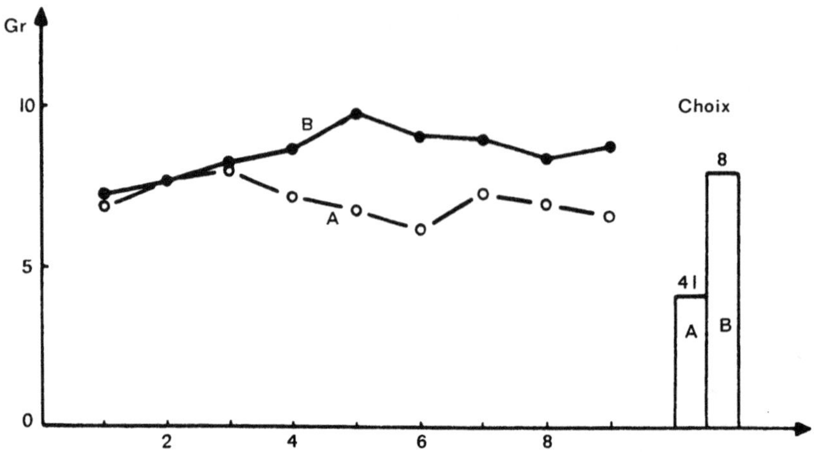

Fig. 7. A flavored diet, supplemented after a lh daily meal by a postprandial administration of a glucose solution, (*A*) is progressively less consumed than the same differently flavored and nonsupplemented diet alternately presented. (*B*) After 9 days, rats exhibit in a choice their acquired preference for the previously nonsupplemented and less satiating diet (Le Magnen 1959)

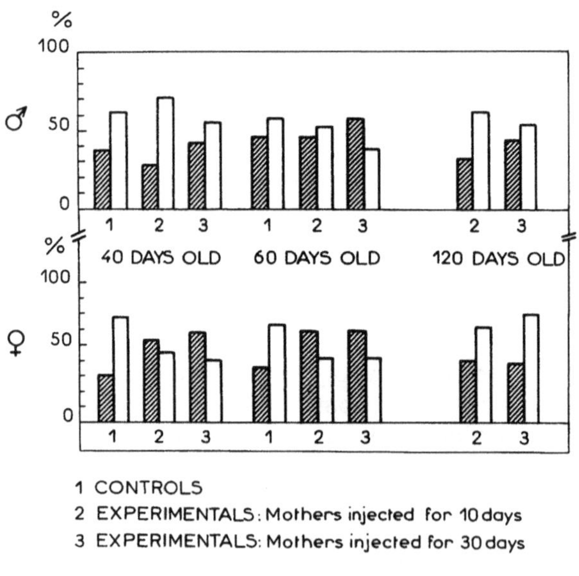

Fig. 8. Preference of young rats (tested at 40, 60 and 120 days of age) for a diet flavored by adding an odorous compound injected to their lactating mothers for either 10 or 30 days of suckling. (Le Magnen and Tallon 1968)

The same conditioned food-odor preference has been demonstrated during the suckling period. Mother lactating rats were injected for 10 days with a minute amount of an odorant (citral) so that the young were fed with flavored mother milk. After weaning these pups, compared with controls, exhibited a preference for food labeled with the same odorant (Le Magnen 1968) (Fig. 8). Later it was demonstrated that pups acquired a preference of the mother diet through the milk, in which there were traces of odorants from the diet (Galef and Osborne 1978; Galef and Sherry 1973).

5.2 Learned Central Processing of the Olfactory Input to the Brain

Similar to the learned modification of taste discrimination, the learned responses to food odors imply a learned central processing of the olfactory input into the brain. This effect of learning on the olfactory bulb electrical responses was suggested for the first time by Goldberg and Moulton (1969). Later a series of works showed in rats the hunger-dependency of the multiunit response to food odor recorded in the mitral cell layer (Giachetti et al. 1972; Pager et al. 1972; Pager 1974) (Fig. 9). Discharges under stimulation by the odor of the familiar food – and only by this food odor – were enhanced in hungry rats compared with satiated animals. These hunger-dependent responses and their specificity to the food odor are learned. Any odorant added to the stock diet after weaning elicits in adulthood a hunger-dependent response in the olfactory bulb in contrast to results in controls. Inflating a balloon in the stomach, like the oral intake of the food, abolishes this differential response to food-related and not related odors (Chaput and Holley 1976). This modulation by hunger and satiety of the olfactory bulb response to the food odor disappears after a transection of centrifugal pathways to the bulb (Pager 1974).
This selective modulation of olfactory bulb activity suggests another aspect of the central processing of olfactory information. The identification of a mixture of chemicals as the odor of a particular food object implies that the olfactory system is able to extract, and the brain to recognize, the complex message generated by this mixture among the noise and other ambient odors. The odor of coffee for example, which is a mixture of more than 500 different pure chemicals, is iden-

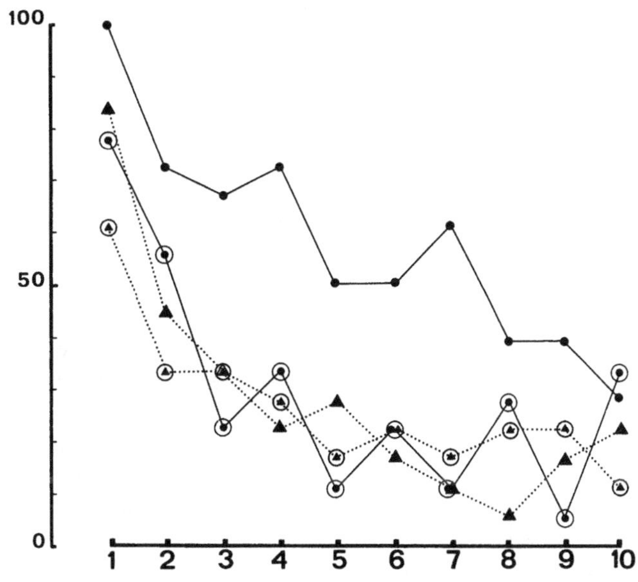

Fig. 9. Percentage of multiunit positive responses in the mitral cell layer of the olfactory bulb to 10 successive puffs of stimulation by the food odor (● and ⊙) or amylacetate (▲ and Ⓐ) in rats either hungry (● and ▲) or satiated (⊙ and Ⓐ). (Prager 1978)

tified among other ambient odors. Such a processing is common in hearing and vision. The odor of coffee is recognized just as the complex mixture of sound frequencies representing the tone of a violin is recognized and is separated from the other tones in an orchestra. This suggests that, at some level of the sensory input, the across-fiber pattern specific to the known mixture of pure stimuli is segregated and separated from – and not added to – the other simultaneous messages. This important point has hardly been investigated in olfaction. However, early experiments showed that the integrated responses of olfactory bulbs in cats were rapidly habituated when stimulated by biologically insignificant odors. The responses to a fishy odor were not habituated, and induced a strong increase of activity in the hungry animal (Hernandez-Péon et al. 1961). A more recent report also suggests that this process is localized in the olfactory bulb (Potter and Chorover 1976). The adaptation to continuous odor stimulation was simultaneously recorded by the electro-olfactogram at the level of the olfactory mucosa and of the mitral cells in the bulbs. Upon interruption of the stimulation, the response was rapidly recovered at the level of olfactory mucosa while the unresponsiveness persisted and was only recovered slowly in the mitral cells. This slow recovery is interpreted by the author as habituation. However, he fails to demonstrate the role of centrifugal input to the bulb in this presumed habituation. The differential habituation of the olfactory bulb response to a food-related mixture of stimuli, compared with associated irrelevant odors, has still not been demonstrated.

6 Associated Sensory Cues and the Best Channel in Palatability Learning

Olfactory and gustatory cues, and their respective role in feeding, are studied separately in experimental conditions. This separation is rarely achieved in common foods and beverages. Most of them generate a complex pattern of visual, chemical, mechanical, and thermic stimulations outside and inside the mouth. Palatability responses are based upon this complex pattern.

The role of food texture and therefore of tactile and proprioceptive sensory systems in the oral cavity has been largely neglected by experimenters. In rats, greasy texture seems to be an unlearned sensory stimulus to eat (Hamilton 1964; Tarozzi et al. 1977; Martin and Bellingham 1979; Galili et al. 1981). The intake of a familiar food is augmented by the addition of a nutritive or nonnutritive greasy material. The palatability due to the nonnutritive material (vaseline, for example) is rapidly extinguished (Hamilton 1964) as can be observed with artificial sweeteners. Rats give different responses to their stock diet presented as biscuit or in a powdered form (Corbit and Stellar 1964). When satiated by the intake of a sucrose solution, they renew their intake when presented with crystallized sucrose (Mook et al. 1983). The role of consistency in human likes and dislikes is part of common experience.

Various sensory channels, separately or in association, possess ability to serve palatability learning. The channels that best subserve feeding have been studied particularly by using conditioned taste aversion. These differences were tested experimentally in rats with cues other than taste. Two forms of the same diet, one containing amphetamine as UCS, were presented alternately. In different groups, the two forms were made discriminable either by adding different odors, by different textures (powdered and crystallized sugars), by colors, or by fixed positions in the cage (spatial cue). A final choice, which displayed the acquired aversion to the form previously associated to amphetamine, gave the following order of the strength of aversion: odor, texture, spatial cues, color (Le Magnen 1956). In rodents, taste is the best channel. Pairing taste cues with postingestive toxicosis only produces the one-trial and long-delay learning of aversion. Olfactory cues are less effective (Simson and Booth 1973). A potentiation of olfactory stimuli by their association with taste has been recognized. In rats, the association of an odor cue to a novel taste potentiates the induction of odor aversion (Rusiniak et al. 1979, 1982a, b). Convergences of olfactory and gustatory pathways in the NTS and the thalamus are presumably a neuroanatomical and functional basis of these interactions. Food conditioning based upon visual cues is also difficult in rodents and is potentiated by association with taste (Berk and Miller 1978; Galef and Osborne 1978).

These best channels for conditioning responses to food vary according to species and (it is an important point) according to the prominent role apparently played in feeding behavior in each animal family. In nonhuman primates, conditioning using visual cues is readily obtained (Johnson et al. 1975). This was assessed in the neurophysiological experiments reported above, in which the electrical discharges of an hypothalamic single unit was shown to be conditioned by the monkey's viewing of a food object (Mora et al. 1976). In birds, whose feeding is guided more visually than chemically, pairing visual cues with toxicosis induces aversion more effectively than pairing with taste cues (Martin and Bellingham 1979; Westbrook et al. 1980).

7 Brain Mechanisms of Behaviors, Learning and Reward and Their Relationships

The discussion of the data examined in the preceding sections leads to the following statements: Some sensory stimuli from foods are state-dependent, unconditioned stimuli to eat. Most of them become conditioned stimuli to eat a food and a given amount of that food. In the course of this learning, external and oropharyngeal activities of foods acquire their palatability level, which can also be termed a hedonic value or a sensory-rewarding activity. In this way, animals can anticipate quantitatively and qualitatively postingestive nutritive activities of foods from their sensory activities at the entrance to and within the alimentary canal.

What are the brain mechanisms of learning and of reinforcement? What are the relationships between them and between them and the neuronal circuitry involved in behaviors? This is the decisive final question. It is also the most embarrassing question, since the answer is practically unknown. It has been subject to divergent approaches and conflicting terminologies, which are likely responsible for misunderstanding and "byzantine" discussions. This gap in our knowledge of the essential function of the CNS and even living organisms is indeed a dramatic failure of modern physiology (Van Vort and Smith 1983).

The positive primary reinforcement in the learning of feeding responses is the recovery from hunger, i.e., satiation, or the recovery from nutritive deficiencies as sources of metabolic disorders and illness (Green and Garcia 1971; Parker et al. 1973; Seward and Greathouse 1973; Zahoric et al. 1974). Many experiments (some of them described above) have demonstrated that food loaded in the stomach or parenterally administered, contingent to the oral intake, reinforces the response to the food. The recovery from an induced illness has been shown to be a positive reinforcer (Hasegawa 1981). In all behaviors, the satiation by the behavioral performance, that is to say the relief of the state which has initiated this performance (hunger, thirst, fear, sexual arousal), is the source of an appetitive learning of this performance. Conversely, illness, a postingestive metabolic imbalance, or the lack of satiation contingent to eating a food induces a passive or active avoidance of the food. This may also be extended to all other behaviors.

A first conclusion may be drawn from these facts. Since learning is intrinsically linked to specific behavioral arousal and its satiation, it is necessary to think that the particular neuronal circuitry involved in each behavior (here hunger and satiety) is also the brain mechanism for learning that behavior. Learning and memory have for centuries been conceived as an unique and specific function of the brain which is more or less separate from other brain functions. It was localized somewhere, e.g., in the cortex (by Pavlov), in the hippocampus, etc., and this unique and centralized system was reputed to act through undefined pathways on the various sites and neuronal systems underlying the different behaviors. The above analysis suggests, however, that learning is a neuronal process incorporated into neuronal circuitries and synaptic transmissions involved separately in different behaviors.

Inasmuch as each specific behavior involves a specific network of internal and external inputs and different motor programs, each behavior possesses its own amnesic system in this machinery. It is possible to speculate that feeding, drinking, and defensive memories, which are dependent on the same basic neuronal mechanism, are topographically separate in the same way as the behavioral systems. Anorexia nervosa resembles a partial feeding amnesia and some libido losses a partial sexual amnesia.

Some experimental facts support this notion. Lesioning of the basolateral nuclei of amygdala, which are involved in the feeding circuitry, abolishes the capacity of rats to learn taste preferences and conditioned taste aversions (Rolls and Rolls 1973; Nachman and Ashe 1974; Fitzgerald and Burton 1981; Kolakowska 1984). Lesioning the lateral hypothalamus (critically involved in feeding) also abolishes learned taste aversions (Roth et al. 1973; Schwartz and Teitelbaum 1974). A lesion of the area postrema prevents a conditioning of aversion by scopolamine, but not by amphetamine (Ashe and Nachman 1980). This suggests that different neuronal systems are involved in the learning of palatability and of food rejections, since amphetamine seems to act in relation to its anorexigenic effects and not as inducing illness or malaise (Le Magnen 1969). An ablation of the gustatory neocortex alters the conditioning of a taste aversion; the number of required trials is augmented. The specificity and persistency of the acquired response are diminished (Buresova and Bures 1973; Lorden 1976; Kiefer and Braun 1979).

A second conclusion that may be drawn is that the sensory food reward is also intrinsically linked both to the learning process and to mechanisms underlying the behavior.

In an enormous body of works, this learning or conditioning (the two notions are arbitrarily separated) and its consequences on behavior have been studied by assessing a learned operant or instrumental response (pressing a lever) instead of the behavioral response itself (eating or not eating the food). The food delivered by pressing the lever is said to be "a food reward" of the operant, but many experiments have demonstrated that food is a reward for this motor response, imposed by the sensory and not by the postingestive activity of that food in order to obtain and to eat the pellet of food (Guttman 1953; Wike and Casey 1954). The effect of conditioning and learning is to create a sensory secondary reinforcement, a sensory food reward. Learning is the transfer to external sensory cues of the capacity to stimulate and/or inhibit the current performance.

Brain rewarding systems have been hypothesized and studied as responsible for the reinforcement of the intracranial electrical self-stimulation (ICSS). These brain rewarding systems have been conceived as an unique system distributed in various parts of the brain. Long-lasting controversies have surrounded their unique neurochemical basis; this is noradrenergic for one group of investigators (Wise 1978), dopaminergic for others (Wise and Schwartz 1981; Wise 1982), and opioergic for a third group (Esposito and Kornetsky 1978). A possible relation with the rewarding or reinforcing mechanisms of natural behaviors has been suggested.

Contrary to this unitary notion, many experiments showed that the ICSS in some sites was dependent on a specific behavioral arousal and/or on the presence of sensory stimuli involved in the behavior. LH self-stimulation is hunger dependent, since rats press a lever to escape an imposed electrical stimulation when satiated (Hoebel 1968). External stimulation by the familiar food facilitates this LH self-stimulation (Coons and Cruce 1968). Other data showed the electrical stimulation of LH reinforced a flavor preference as does the postingestive action of the ingested food. Pairing the intake of saccharin with LH electrical stimulation increases the palatability of saccharin. Thus, LH stimulation appears to be a substitute for postingestive reinforcement absent with a saccharin solution and provided by a glucose solution (Ettenberg and White 1978; Ettenberg 1979; Ettenberg et al. 1982). In another experiment, selected lines of rats were separated as high and low LH self-stimulators. High stimulator-selected rats spontaneously drank more of a saccharin sweet solution than did low stimulators (Ganchrow et al. 1981).

Endogenous opiates which are presumably linked with dopaminergic neuronal systems are involved in the sensory food reward. Naloxone, the specific opiate antagonist, and neuroleptic drugs, such as DOPA agonists, abolish the palatability of a sweet solution (Le Magnen et al. 1980b; Xenakis and Sclafani 1981, 1982; Rockwood and Reid 1982; Wise 1982). It introduces, as designated by one of the authors, an "anhedonia." In rats, the intake of a saccharin solution induces a depletion of the endorphin content of hypothalamus (Dum et al. 1983). Also in rats, the intake of a meal – inasmuch as the offered food is palatable – induces as postprandial analgesia which assesses an induced-endorphin release (Louis-Sylvestre et al. submitted).

These last facts suggest that there are behaviorally specific rewarding systems that can be separated neither from the brain system involved in the particular behavior, nor from the learning process relevant to this behavior. Animals memorize what is good or what is bad, and not what is unserviceable. The facilitatory and inhibitory hedonic dimension appears to be incorporated into the neuronal or transneuronal transformation involved in the storage and retrieval of information. Memory is not somewhere in the brain, the brain-rewarding system somewhere else, and behaviors elsewhere again. A specific brain mechanism that is involved in the integration of somatic and sensory information underlying each behavior includes its own learning and rewarding system basically associated with the neuronal activity.

References

Aguilar-Baturoni HU, Guevara-Aguilar R, Arechiga H, Alcocer-Cuaron C (1976) Hypothalamic influences on the electrical activity of the olfactory pathway. Brain Res Bull 1:263–272

Aleksanyan C, Buresova O, Bures J (1976) Modification of unit responses to gustatory stimuli by conditioned taste aversion in rats. Physiol Behav 17:173–179

Aravich PF, Sclafani A (1980) Dietary preference in rats fed bitter tasting quinine and sucrose octa acetate adulterated diets. Physiol Behav 25:157–160

Ardisson JL, Dolisi C, Ozon C, Crenesse D (1981) Caractéristiques des prises d'eau et d'aliments spontanées chez des chiens en situation ad lib. Physiol Behav 26 (3):361–371

Armstrong S (1980) A chronometric approach to the study of feeding behavior. Neurosci Biobehav Rev 4:27–53

Ashe J, Nachman M (1980) Neural mechanisms in taste aversion learning. In: Sprague A, Epstein A (eds) Progress in psychobiology and physiological psychology, vol. 9. Academic, New York, pp 233–262

Balagura S, Coscina DV (1968) Periodicity of food intake in the rat as measured by an operant response. Physiol Behav 3:641–643

Balagura S, Kanner M, Harrell LE (1975) Modification of feeding patterns by glucodynamic hormones. Behav Neurol Biol 13:457–460

Beach FA (1940) Effects of cortical lesions upon the copulatory behavior of male rats. J Comp Psychol 29:193–244

Bellisle F (1979) Human feeding behavior. Neurosci Biobehav Rev 3:163–169

Bellisle F (1984) La stimulation à manger chez le sujet humain: étude des effets des facteurs sensoriels et systémiques sur la microstructure des repas. Thesis, Paris

Bellisle F, Lucas F, Amrani R, Le Magnen J (1984) Deprivation, palatability and the microstructure of meals in human subjects. Appetite 5:85–94

Bereiter DA, Berthoud HR, Jeanrenaud B (1980) Hypothalamic input to brain stem neurons responsive to oropharyngeal stimulation. Exp Brain Res 39:33–39

Berk AM, Miller RR (1978) LiCl-induced aversions to audiovisual cues as a function of response measure and CS-US interval. Behav Neurol Biol 24:185–208

Bernstein IL (1976) Spontaneous activity and meal patterns. In: Novin D, Wyrwicka W, Bray GA (eds) Basic mechanisms and clinical implications. Raven, New York, pp 219–224

Berridge K, Flynn E, Shulkin J, Grill H (1984) Sodium depletion enhances salt palatability in rats. Behav Neurosci 98:652–660

Blass EM, Teicher MH, Cramer CP, Bruno JP, Hall WG (1977) Olfactory, thermal and tactile controls of suckling in preauditory and previsual rats. J Comp Physiol Psychol 91:1248–1260

Booth DA (1972) Conditioned satiety in rats. J Comp Physiol Psychol 81:457–471

Booth DA, Mather P, Fuller J (1982) Starch content of associatively conditions human appetite and satiation, indexed by intake and eating pleasantness of starch-paired flavours. Appetite 3:163–184

Buresova O, Bures J (1973) Cortical and subcortical components of the conditioned saccharin aversion. Physiol Behav 11:435–439

Burghardt GM (1970) Chemical perception in reptiles. In: Johnston JW Jr, Moulton DG, Turk A (eds) Communication by chemical signals. Appleton Century Crofts, New York, pp 214–308

Burton MJ, Mora F, Rolls ET (1975) Visual and taste neurones in the lateral hypothalamus and substantia innominata: modulation of responsiveness by hunger (abstr). J Physiol (Lond) 252:50P

Burton MJ, Rolls ET, Mora F (1977) Effects of hunger on the responses of neurones in the hypothalamus to the sight and taste of food. Exp Neurol 53:508–519

Campfield LA, Brandon P, Smith F (1985) On-line continuous measurement of blood glucose and meal pattern in free-feeding rats: the role of glucose in meal initiation. Brain Res Bull (in press)

Capretta PJ (1964) Saccharin consumption and the reinforcement issue. J Comp Physiol Psychol 57:448–450

Carper JW, Polliard F (1953) A comparison of the intake of glucose and saccharin solutions under conditions of caloric needs. Am J Physiol 66:479–482

Chang FC, Scott TR (1984) Conditioned taste aversions modify neural responses in the rat nucleus tractus solitarius. J Neurosci 4:1850–1862

Chaput M, Holley A (1976) Olfactory bulb response to food odour during stomach distension in the rat. Chem Senses Flavour 2:189–201

Coons EE, Cruce JAF (1968) Lateral hypothalamus: food current intensity and maintaining self-stimulation of hunger. Science 159:1117–1119

Corbit JD, Stellar E (1964) Palatability, food intake and obesity in normal and hyperphagic rats. J Comp Physiol Psychol 58:63–67

Davies RF (1977) Long-short term regulation of feeding patterns in the rat. J Comp Physiol Psychol 91:574–585

De Castro JM (1975) Meal pattern correlations: facts and artifacts. Physiol Behav 15:13–17

Desor JA, Maller O, Turner R (1973) Taste in acceptance of sugars by human infants. J Comp Physiol Psychol 84:496–501

Di Lorenzo PM, Schwartzbaum JS (1980) Coding of taste information in the pontine taste area of the rabbit. In: Van der Starre (ed) Olfaction and taste VII. IRL, London, pp 251–254

Di Lorenzo PM, Schwartzbaum JS (1982) Coding of gustatory information in the pontine parabrachial nuclei of the rabbit. Brain Res 251:229–258

Doetsch G, Erickson R (1970) Synaptic processing of taste quality information in the NTS of the rat. J Neurophysiol 33:490–507

Dugas du Villard X, Her C, MacLéod P (1981) Qualitative discrimination of sweet stimuli: behavioral study on rats. Chem Senses 6:143–148

Dum J, Gramsch CH, Herz A (1983) Activation of hypothalamic B-endorphin pools by reward induced by highly palatable food. Pharmacol Biochem Behav 18:443–447

Duncan IJH, Horne AR, Hughes BO, Wood-Gush DGM (1970) The pattern of food intake in female brown leghorn fowls as recorded in a skinner box. Anim Behav 18:245–255

Erickson RP (1963) Sensory neural patterns and gustation. In: Zotterman Y (ed) Olfaction and taste. Pergamon, Oxford, p 205

Erickson RP, Gernot S, Doetsch G, Marshall DA (1965) The gustatory neural response function. J Gen Physiol 49:247–263

Erulkar SD (1972) Comparative aspects of spatial localization of sound. Physiol Rev 52:237–360

Esposito RU, Kornetsky C (1978) Opioids and rewarding brain stimulation. Neurosci Biobehav Rev 2:115–122

Ettenberg A (1979) Conditioned taste preferences as a measure of brain-stimulation reward in rats. Physiol Behav 23:167–172

Ettenberg A, White N (1978) Conditioned taste preferences in the rat induced by self-stimulation. Physiol Behav 21:363–368

Ettenberg A, Sgro S, White N (1982) Algebraic summation of the affective properties of a rewarding and an aversive stimulus in the rat. Physiol Behav 28:873–878

Faurion A (1982) Etudes des mécanismes de la chemioreception du goût sucré. Thesis, Paris

Fitzgerald RE, Burton MJ (1981) Effects of small basolateral amygdala lesions on ingestion in the rat. Physiol Behav 27:431–437

Frank M (1977) The distinctiveness of responses to sweet in the chorda tympani nerve. In: Weiffenbach JM (ed) Taste and development: the genesis of sweet preference. US Dept Health, Education and Welfare, Bethesda 25–39

Frank M, Pfaffmann C (1969) Taste nerve fibers: a random distribution of sensitivities to four tastes. Science 164:1183–1185

Galef BG, Henderson PW (1972) Mother's milk: a determinant of the feeding preferences of weanling rat pups. J Comp Physiol Psychol 78:213–219

Galef BG, Osborne B (1978) Novel taste facilitation of the association of visual cues with toxicosis in rats. J Comp Physiol Psychol 92:907–916

Galef BG, Sherry DF (1973) Mother's milk: a medium for transmission of cues reflecting the flavor of mother's diet. J Comp Physiol Psychol 83:374–378

Galili D, Maller O, Brightman VJ (1981) The effect of desalivation by duct ligation or salivary gland extirpation on taste preference in rats. Arch Oral Biol 26:853–858

Ganchrow JR, Erickson RP (1970) Neural correlates of gustatory intensity and quality. J Neurophysiol 33:768–783

Ganchrow JR, Lieblich I, Cohen E (1981) Consummatory responses to taste stimuli in rats selected for high and low rates of self stimulation. Physiol Behav 27:971–976

Geary N, Langhans W, Scherrer E (1981) Metabolic concomitants of gucagon-induced suppression of feeding in the rat. Am J Physiol 241:R 330–335

Giachetti I, MacLéod P, Le Magnen J (1972) Contrôle centrifuge des afférences olfactives en fonction des états de faim et de satiété chez le rat. CR Soc Biol (Paris) 164:841

Giza BK, Scott TR (1983) Blood glucose selectively affects taste-evoked activity in rat nucleus tractus solitarius. Physiol Behav 31:643–650

Glenn JF, Erickson RP (1976) Gastric modulation of gustatory afferent activity. Physiol Behav 16:561–568

Goldberg SJ, Moulton DG (1969) Multi- and single-unit activity telemetered from the olfactory bulb of rats involved in an odor detection task. Am Zool 9:593

Green KF, Garcia J (1971) Recuperation from illness: flavor enhancement for rats. Science 173:749–751

Grill HJ, Norgren R (1978) Chronically decerebrate rats demonstrate satiation but not bait shyness. Science 201:267–269

Guttman N (1953) Operant conditioning, extinction and periodic reinforcement in relation to concentration of sucrose used as a reinforcing agent. J Exp Psychol 46:231–224

Halpern BP (1965) Chemotopic organization in the bulbar gustatory relay of the rat. Nature 208:5008, 393–395

Hamilton (1964) Rat's preference for high fat diets. J Comp Physiol Psychol 58:459–460

Hasegawa Y (1981) Recuperation from lithium-induced illness: flavor enhancement for rats. Behav Neurol Biol 33:252–255

Harris LJ, Clay J, Hargreaves FJ, Ward A (1933) Appetite and choice of diet: the ability of the vitamin B deficient rat to discriminate between diets containing and lacking the vitamin. Proc R Soc Lond [Biol] 113:161–190

Haybach JP, Boyle PC (1982) Dietary quinine reduces body weight and food intake independent of aversive taste. Physiol Behav 29:1171–1174

Hernandez-Peon R, Brust-Carmona H, Penaloza-Rojas J, Bach-Y-Rita G (1961) The efferent control of afferent signal entering the central nervous system. Ann NY Acad Sci 89:866–882

Hoebel BG (1968) Inhibition and disinhibition of self-stimulation and feeding: hypothalamic control and postingestional factors. J Comp Physiol Psychol 66:89–100

Holley A, MacLéod P (1977) Transduction et codage des informations olfactives chez les vertébrés. J Physiol (Paris) 73:725–828

Hsiao S, Tuntland P (1971) Short-term satiety signals generated by saccharin and glucose solutions. Physiol Behav 7:287–289

Jacobs HL (1962) Some physical, metabolic and sensory components in the appetite for glucose. Am J Physiol 203:1043–1054

Johnson C, Beaton R, Hall K (1975) Poison-based avoidance learning in nonhuman primates: use of visual cues. Physiol Behav 14:403–408

Keidel WD, Neff WD (eds) (1974) Auditory system (Handbook of sensory physiology, vol 5). Springer, Berlin Heidelberg New York, pp 736

Kiefer SW, Braun JJ (1979) Acquisition of taste avoidance habits in rats lacking gustatory neocortex. Physiol Psychol 7:245–250

Kissileff HR (1970) Free feeding in normal and "recovered lateral" rats monitored by a pellet-detecting eatometer. Physiol Behav 5:163–174

Kolakowska L (1984) Effet des lésions de l'amygdale sur la manifestation et l'apprentissage des réponses ingestives chez le rat. Thesis, Paris

Kratz CM, Levitsky DA, Lustick SL (1978a) Differential effects of quinine and sucrose octoacetate on food intake in the rat. Physiol Behav 20:665–668

Kratz CM, Levitsky DA, Lustick SL (1978b) Long term effects of quinine on food intake and body weight in the rat. Physiol Behav 21:321–324

Lagaillarde M, Louis-Sylvestre J (1986) Delaying the preprandial fall in blood glucose level delays spontaneous meal onset (unpublished)

Lancet D, Greer CH, Kauer G, Shepherd G (1982) Mapping of odor-related neuronal activity in the olfactory bulb by high-resolution 2-deoxy glucose autoradiography. Proc Nat Acad Sci USA 79:670–674

Larue-Achagiotis C, Le Magnen J (1985a) Effect of long-term insulin infusion on body weight and food intake: intravenous versus intraperitoneal routes. Appetite 6:319–329

Larue-Achagiotis C, Le Magnen J (1985b) Feeding rates and responses to food deprivation as a function of fast-induced hypoglycemia. To be published

Lawless H (1979) Evidence for neural inhibition in bittersweet taste mixtures. J Comp Physiol Psychol 93:538–547

Le Magnen J (1953) Activité de l'insuline sur la consommation spontanée de solutions sapides. CR Soc Biol (Paris) 147:1753–1757

Le Magnen J (1954) Le processus de discrimination par le rat blanc des stimuli sucrés alimentaires et non alimentaires. J Physiol (Paris) 46:414–418

Le Magnen J (1956) Hyperphagie provoquée chez le rat blanc par altération du mécanisme de satiété périphérique. CR Soc Biol (Paris) 150:32–34

Le Magnen J (1959) Effets des administrations post-prandiales de glucose sur l'etablissement des appétits. CR Soc Biol (Paris) 153:212–215

Le Magnen J (1960) Effets d'une pluralité de stimuli alimentaires sur le déterminisme quantitatif de l'ingestion chez le rat blanc. Arch Sci Physiol 14:411–419

Le Magnen J (1968) Préférence alimentaire du jeune rat induite par l'allaitement maternel. CR Soc Biol (Paris) 162:387–390

Le Magnen J, Devos M (1970) Metabolic correlates of the meal onset in the free food intake of rats. Physiol Behav 5:805–814

Le Magnen J, Devos M (1980) Parameters of the meal pattern in rats: their assessment and physiological significance. Neurosci Biobehav Rev 4 (Suppl 1):1–11
Le Magnen J, Devos M (1982) Daily body energy balance in rats. Physiol Behav 29:807–811
Le Magnen J, Devos M (1984) Meal to meal energy balance. Physiol Behav 32:39
Le Magnen J, Louis-Sylvestre J (1980) Palatability and preabsorptive insulin release. Neurosci Biobehav Rev 4 (Suppl 1):43–46
Le Magnen J, Tallon S (1968) Préférence alimentaire du jeune rat induite par l'allaitement maternel. CR Soc Biol (Paris), 162:387–389
Le Magnen J, Tallon S (1963) Enregistrement et analyse préliminaire de la périodicité alimentaire spontanée chez le rat blanc. J Physiol (Paris) 55:286–297
Le Magnen J, Tallon S (1966) La périodicité spontanée de la prise d'aliments ad libitum du rat blanc. J Physiol (Paris) 58:323–349
Le Magnen J, Devos M, Gaudillière JP, Louis-Sylvestre J, Tallon S (1973) Role of a lipostatic mechanism in regulation by feeding of energy balance in rats. J Comp Physiol Psychol 84:1–23
Le Magnen J, Devos M, Larue-Achagiotis C (1980a) Food deprivation induced parallel changes in blood glucose, plasma free fatty acids and feeding during the two parts of the diurnal cycle in rats. Neurosci Biobehav Rev 4 (Suppl 1):17–23
Le Magnen J (1969) Peripheral and systemic actions of food in the caloric regulation of intake. Ann NY Acad Sci 157 (2):1126–1127
Le Magnen J (1975) Olfactory endocrine interactions and regulatory behaviour. In: Denton DA, Coghlan JP (eds) Olfaction and taste V/Academic, New York. pp 381–388
Le Magnen J (1976) Interactions of glucostatic and lipostatic mechanisms in the regulatory control of feeding. In: Novin D, Wyrwicka W, Bray GA (eds) Hunger: basic mechanisms and clinical implication. Raven, New York, pp 89–101
Le Magnen J (1977a) Sweet preference and the sensory control of caloric intake. In: Weiffenbach J (ed) Taste and development. US Dept of Health, Education and Welfare, Bethesda, pp 355–362
Le Magnen J (1977b) Hunger and food palatability in the control of feeding behaviour. In: Katsuki Y, Sato M, Tagaki S, Oomura Y (eds) Food intake and chemical senses. University of Tokyo, Press Tokyo, pp 263–280
Le Magnen, J (1981) The metabolic basis of dual periodicity of feeding in rats. Behav Brain Sci 4:561–607
Le Magnen J (1983) Body energy balance and food intake: a neuroendocrine regulatory mechanism. Physiol Rev 63:314–386
Le Magnen J (1984) Is regulation of body weight elucidated? Neurosci Biobehav Rev 8:515–522
Le Magnen J (1987) Palatability: concepts, terminology and mechanisms. In: Boakes RA, Popplewell DA, Burton MJ (eds) Eating habits. Wiley, New York
Le Magnen J, Marfaing-Jallat P, Miceli D, Devos M (1980b) Pain modulating and reward systems: a single brain mechanism? Pharmacol Biochem Behav 12:729–733
Lorden, JF (1976) Effects of lesions of the gustatory neocortex on taste aversion learning in the rat. J Comp Physiol Psychol 90:665–679
Louis-Sylvestre J (1976) Pre-absorptive insulin release and hypoglycemia in rats. Am J Physiol 230:56
Louis-Sylvestre J, Le Magnen J (1980) A fall in blood glucose level precedes meal onset in free-feeding rats. Neurosci-Biobehav-Rev 4 (Suppl 1):13–15
Louis-Sylvestre J (1984) Méchanismes de l'induction de l'hyperphagie et de l'obésité par le régime caféteria: hypothéses? Cah Nutr Diet XIX:197–204
Marcilloux JC (1980) Comportement alimentaire de l'oie de race landaise: importance du noyau ventromédian de l'hypothalamus dans sa régulation. Thesis, Paris

Marg E (1964) The accessory optic system. Ann NY Acad Sci 117:35–52

Martin GM, Bellingham WP (1979) Learning of visual food aversions by chickens (Gallus gallus) over long delays. Behav Neurol Biol 25:58–68

Matsuo Y, Shimazu Y, Jusano K (1984) LH modulation of oral different activity in NTS neurons of rats. J Neurosci 4:1201–1207

McCutcheon B, Brown L (1983) Response to NaCl taste in mixture with sucrose by sodium deficient rats. Physiol Behav 30:405–408

Mei N (1982) Sensory structure in the viscera. In: Autrum H, Ottoson D, Perl ER, Schmidt RF, Shimazu H, Willis WD (eds) Progress in sensory physiology, vol 4, Springer, Berlin Heidelberg New York, pp 1–42

Mook DG, Brane JA, Whitt JA (1983) De-satiation: the reinstatement of feeding in glucose-satiated rats. Appetite 4:125–136

Mora F, Rolls ET, Burton MJ (1976) Modulation during learning of the responses of neurons in the lateral hypothalamus to the sight of food. Exp Neurol 53:508–519

Mrosowsky N, Sherry DF (1980) Animal anorexias. Science 207:837–842

Nachman M (1963) Learned aversion to the taste of lithium chloride and generalization to other salts. J Comp Physiol Psychol 56:343–349

Nachman M, Ashe JH (1974) Effects of basolateral amygdala lesions on neophobia, learned taste aversion and sodium appetite in rats. J Comp Physiol Psychol 87:622–643

Natelson BH, Bonbright JC (1978) Patterns of eating and drinking in the monkeys when food and water are free and when they are earned. Physiol Behav 21:201–214

Norgren R (1970) Gustatory responses in the hypothalamus. Brain Res 21:63–77

Norgren R (1976) Taste pathways to hypothalamus and amygdala. J Comp Neurol 166:17–30

Norgren R (1977) A synopsis of gustatory neuroanatomy. In: Le Magnen J, Mac Léod P (eds). Olfaction and taste, vol VI. IRL, London, pp 225–232

Norgren R (1981) The central organization of gustatory and visceral afferent system in the NTS. In: Katsuki Y, Norgren R, Sato M (eds) Brain mechanisms of sensation, Wiley, New York, pp 143–160

Nowlis GH, Frank ME, Pfaffmann C (1980) Specificity of acquired aversions to taste qualities in hamsters and rats. J Comp Physiol Psychol 94:932–942

Odum EP (1960) Premigratory hyperphagia in birds. Am J Clin Nutr 8:621–629

Pager J (1974) A selective modulation of olfactory input suppressed by lesions of the anterior limb of the anterior commissure. Physiol Behav 13:523–526

Pager J, Giachetti I, Holley A, Le Magnen J (1972) A selective control of olfactory bulb electrical activity in relation to food deprivation and satiety in rats. Physiol Behav 9:573–580

Pager J (1978) Ascending olfactory information and centrifugal influxes contributing to a nutritional modulation of the rat mitcal cell responses. Br Res 140:251–259

Parker L, Failor A, Weidman K (1973) Conditioned preferences in the rat with a unnatural need state: morphine withdrawal. J Comp Physiol Psychol 82:294–300

Pfaffmann C (1936) Differential responses of the newborn cat to gustatory stimuli. J. Genet Psychol 49:61–67

Pfaffmann C (1974) The sensory coding of taste quality. Chem Senses Flavor 1:5–8

Pfaffmann C, Bare JK (1950) Gustatory nerve discharges on normal and adrenalectomized rats. J Comp Physiol 43:329–334

Pfaffmann C, Hagström EC (1955) Factors influencing taste sensitivity to sugars. Am J Physiol 183:651

Pierson A, Le Magnen J (1969) Etude quantitative du processus de régulation des réponses alimentaires chez l'homme. Physiol Behav 4:61–67

Potter H, Chorover SL (1976) Response plasticity in hamster olfactory bulb: peripheral and central process. Brain Res 116:417–436

Price JL, Powell PS (1970) An experimental study of the origin and the course of the centrifugal fibers to the olfactory bulb in the rat. J Anat 107(2):215–237
Rapuzzi G, Casella C (1963) Studio electrofisiologico sull'innervazions delle papille fungiformi della lingua di rana. Boll Soc Ital Biol Sper 38, 24:1595–1596
Repérant J, Vesselkin NP, Rio JP, Ermakova TV, Miceli D, Peyrichoux J, Weidner C (1981) La voie visuelle centrifuge n'existe-t-elle que chez les oiseaux? Rev Can Biol Exp 40(1):29–46
Ricardo JA, Koh ET (1978) Anatomical evidence of direct projections from the nucleus of the solitary tract to the hypothalamus, amygdala and other forebrain structures in the rat. Brain Res 153:1–26
Richter CP (1936) Increased salt appetite in adrenalectomized rats. Am J Physiol 115:155–161
Riley J, Card J, Mootr R (1981) A retinal projection in the LH in the rat. Cell Tissue Res 214:257–269
Rockwood GA, Reid LD (1982) Naloxone modifies sugar-water intake in rats drinking with open gastric fistulas. Physiol Behav 29(6):1175–1178
Rogers PJ, Blundell JE (1980) Investigation of food selection and meal parameters during the development of dietary-induced obesity. Appetite 1:85
Rogers RC, Kita H, Butcher LL, Novin D (1980) Afferent projections to the dorsal motor nucleus of the vagus. Brain Res Bull 5:365–373
Rolls ET (1972) Activation of amygdaloid neurons in reward eating and drinking elicited by electrical stimulation of the brain. Brain Res 45:365–381
Rolls ET, Rolls BJ (1973) Altered food preferences after lesions in the basolateral region of the amygdala in the rat. J Comp Physiol Psychol 83:248–259
Rolls ET, Burton MJ, Mora F (1976) Hypothalamic neuronal responses associated with the sight of food. Brain Res 111:53–66
Rolls BJ, Rowe EA, Turner R (1980) Persistent obesity in rats following a period of consumption of a mixed high energy diet. J Physiol (Lond) 298:415–428
Roth SR, Schwartz M, Teitelbaum P (1973) Failure of recovered lateral hypothalamic rats to learn specific food aversions. J Comp Physiol Psychol 83:184–197
Rozin P, Kalat JW (1971) Specific hungers and poison avoidance as adaptive specialization of learning. Psychol Rev 78:459–486
Rusiniak KW, Hankins WG, Garcia J, Brett LP (1979) Flavor illness aversions: potentiation of odor by taste in rats. Behav Neurol Biol 25:1–17
Rusiniak KW, Palmerino CC, Garcia J (1982a) Potentiation of odor by taste in rats: tests of some nonassociative factors. J Comp Physiol Psychol 96:775
Rusiniak KW, Palmerino CC, Rice AG, Forthman DL, Garcia J (1982b) Flavor-illness aversions: potentiation of odor by taste with toxin but not shock in rats. J Comp Physiol Psychol 96:527–539
Sanderson JD, Vanderweele DA (1975) Analysis of feeding patterns in normal and vagotomized rabbits. Physiol Behav 15:357–364
Saper CB, Loewy AD, Swanson LW, Cowan WM (1976) Direct hypothalamo-autonomic connections. Brain Res 117:305–312
Schiffman SS, Cahn H, Lindley MG (1981) Multiple receptor site mediate sweetness: evidence from cross adaptation. Pharmacol Biochem Behav 15:377–388
Schwartz M, Scott T (1976) Gustatory neural coding in the rats. Brain Res 130:283–300
Schwartz M, Teitelbaum P (1974) Dissociation between learning and remembering in rats with lesions in the lateral hypothalamus. J Comp Physiol Psychol 87:384–398
Sclafani A, Springer D (1976) Dietary obesity in adult rats: similarities to hypothalamic and human obesity syndromes. Physiol Behav 17:461–471
Scott TR (1974) Behavioral support for a neural taste theory. Physiol Behav 12:413–418
Scott TR (1977) Information processing in the taste system. In: Le Magnen J, Mac Léod P (eds) Olfaction and taste, vol VI. IRL, London, pp 249–256

Scott JW, Chafin BR (1975) Origin of olfactory projections to lateral hypothalamus and nuclei gemini of the rat. Brain Res 88:64–68
Scott JW, Leonard CM (1971) The olfactory connections of the lateral hypothalamus in the rat, mouse and hamster. J Comp Neurol 141:331–344
Scott TR, Perrotto RS (1980) Intensity coding in pontine taste area: gustatory information is processed similarly throughout rats' brain stem. J Neurophysiol 44:739–750
Seward JP, Greathouse SR (1973) Appetitive and aversive conditioning in Thiamine-deficient rats. J Comp Physiol Psychol 83:157–168
Shafa F, Meisami E (1977) A horseradish peroxidase study of the origin of central projections to the rat olfaction bulb. Brain Res 136:355–359
Sharma KN, Jacobs HL, Gopal V, Dua-Sharma S (1977) Nutritional state/taste interactions in food intake: behavioral and physiological evidence for gastric/tate modulation. In: Kare MR, Maller O (eds) Chemical senses and nutrition. Academic, New York, pp 167–188
Shepherd GM (1977) Central processing of olfactory signals. In: Muller-Schwarze D, Mozell MM (eds) Chemical signals in vertebrates. pp 489–497
Shulkin J (1982) Behavior of sodium-deficient rats: the search for the salty taste. J Comp Physiol Psychol 96:628–634
Simson PC, Booth DA (1973) Olfactory conditioning by association with histidine-free or balanced amino acid loads in rats. Q J Exp Psychol 25: 354–359
Smith DV, Travers JB, Van Buskirk RL (1979) Brain stem correlates of gustatory similarity in the hamster. Brain Res Bull 4:359–372
Smotherman WP (1982) Odor aversion learning by the rat fetus. Physiol Behav 29:769–772
Steiner JE (1973) The gusto-facial response: observation on normal and anencephalic newborn infants. In: Bosn JF (ed) Oral sensation and perception: development in the fetus and infant. US Dept of Health, Education and Welfare, Bethesda, pp 254–278
Stickrod G, Kimble DP, Smotherman WP (1982) In utero taste/odor aversion conditioning in the rat. Physiol Behav 28:5–7
Strouthes A (1971) Thirst and saccharin preference in rats. Physiol Behav 6:287–292
Tarozzi G, Di Bella L, Scalera G, Rossi MT (1977) Role of proprioceptive masticatory afferences in the rat feeding pattern behaviour. Nutr Metabol [Suppl] 21:70–72
Teicher MH, Blass EM (1977) First suckling response of the newborn albino rat: the roles of olfaction and amniotic fluid. Science 198:635–637
Teicher MH, Flaum LE, Williams M, Eckhert SJ, Lumia AR (1978) Survival, growth and suckling behavior of neonatally bulbectomized rats. Physiol Behav 21:553–562
Van Buskirk RL, Erickson RP (1977) Odorant responses in taste neurons of the rat NTS. Brain Res 135:287–303
Van Vort W, Smith GP (1983) The relationship between the positive reinforcing and satiating effects of a meal in the rat. Physiol Behav 30:279–284
Westbrook RF, Clarke JC, Provost S (1980) Long-delay learning in the pigeon: flavor, color and flavor-mediated color aversions. Behav Neurol Biol 28:398–407
Wike EL, Casey A (1954) The secondary reward value of food for satiated animals. J Comp Physiol Psychol 47:441–443
Wise RA (1978) Catecholamine theories of reward: a critical review. Brain Res 152:215–248
Wise RA (1982) Neuroleptics and operant behavior: the anhedonia hypothesis. Behav Brain Sci 5:39–52
Wise RA, Schwartz HV (1981) Pimozide attenuates acquisition of lever-pressing for food in rats. Pharmacol Biochem Behav 15:655–656
Wolf LL, Hainsworth FR (1977) Temporal patterning of feeding by humming-birds. Anim Behav 25:976–989

Xenakis S, Sclafani A (1981) The effects of pimozide on the consumption of a palatable saccharin-glucose solution in the rat. Pharmacol Biochem Behav 15:435–442

Xenakis S, Sclafani A (1982) The dopaminergic mediation of a sweet reward in normal and VMH hyperphagic rats. Pharmacol Biochem Behav 16:293–302

Yamamoto T, Matsuo R, Kawamura Y (1980a) Corticofugal effects on the activity of thalamic taste cells. Brain Res 193:258–262

Yamamoto T, Matsuo R, Kawamura Y (1980b) Localization of cortical gustatory area in rats and its role in taste discrimination. J Neurophysiol 44:440–455

Young PT, Falk JL (1956) The relative acceptability of sodium chloride solutions as a function of concentration and water need. J Comp Physiol Psychol 49:569–575

Zahorik DM, Maier SF, Pies RW (1974) Preferences for tastes paired with recovery from thiamine deficiency in rats: appetitive conditioning or learned safety. J Comp Physiol Psychol 87:1083–1091

Physiology of the Sweet Taste

A. Faurion

Laboratoire de Neurobiologie Sensorielle, Ecole Pratique de Hautes Etudes,
1 avenue des Olympiades, F-91305 Massy

Progress in Sensory Physiology. Vol 8

1 Introduction

In this paper we review the present status of the physiology of the sweet taste. This involves the reception of an external stimulus by chemoreceptors (molecules), the sensitivity of neural elements and the qualitative and intensity aspects of perception. However, with regard to some physiological aspects, the sweet taste cannot actually be considered in isolation from other taste qualities. Part A of this paper covers concepts and data relevant to general gustation; Part B is devoted more specifically to the sweet taste.

Data are presented to show that recently stepped-up focusing on sweet taste research leads to a questioning of theories which up to now have been simply accepted and to the conclusion that the physiology of gustation is presently not easily understood. Much simplification in the interpretation of experimental results has occurred in the past decades, which is the way to build satisfactory knowledge; but previous data can sometimes be re-interpreted from new points of view. For example, taste, mouth somesthesis and olfaction, which are the components of food flavours, were properly distinguished by Chevreul in 1824. Since then, biologists have endeavoured to separate them in their experimental studies and interpretative arguments. Although it is undoubtedly wise in experimental work to separate, as far as possible, the non-homogeneous elements of a system, it may not be so in considering the function of the system as a whole, which is qualitatively different from the sum of its elements. We shall cast some doubt upon the concept of a purely gustatory sensory image.

Taste qualities are traditionally divided into four categories: sweet, which is represented by sucrose, bitter by quinine or quinine hydrochloride, salty by NaCl and acid by hydrochloric acid. The work of the psychologist Henning (1916) is most popularly quoted as indisputable support for this division. However, Woolston and Erickson (1979) have pointed out the general misunderstanding of Henning's theory. Henning wrote his paper in opposition to Cohn's (1914) assertion that taste should be described by four isolated qualities. Although agreeing with Cohn's chemical findings concerning the stimulatory part of the molecules, Henning was rather adamant against Cohn's idea concerning the psychological aspect of taste. Henning developed the idea of a taste quality continuum (*kontinuierliche Reihe*) and suggested the surface of a tetrahedron as a graphic display representing the whole set of taste qualities (now called the taste space); the paper also contains a criticism of Kiesow's (1898) analytical model. Thus, Henning proposed his model to account for the range of all different qualities (*einfachen Geschmäcken*) which lie between the basic qualities (*Grundqualitäten*). "Indeed, which layman could possibly confuse sodium chloride, sodium bromide and potassium iodide even if all three are salty?" He insisted upon the fact that all these intermediary qualities are not perceived as compound tastes as if taste would depend on four "isolated" qualities. Although the four qualities must be prominent in his taste model, they are not the basis for a reductive categorization into four qualities but, on the contrary, the starting points to define continuous series of different qualities.

Henning's dissertation is not at all obsolete. Although dealing with psychological perception of quality only and not with receptor site selectivity or neural sensitivity, Henning's expression falls clearly in agreement with the sum of recent physiological data examined in this paper.

The concept of "four basic qualities" appears a very difficult one to deal with and we shall see further that any categorization into four qualities is questionable, from the point of view of the physiologist, even if it holds to some extent from a psychological point of view.

As most neurophysiologists and psychophysicists always used the following four stimuli: sucrose, NaCl, HCl, and quinine as a given experimental set, thus more or less tacitly admitting that they represented "*the* four taste qualities", they biased their interpretations and limited their understanding of taste mechanisms. Let us re-examine and re-interpret these data. In the expression "four basic stimuli", that authors have extensively used, the word "basic" will be taken in its phenomenological sense and we shall often prefer "four reference stimuli" or "standards".

This paper is a compilation of arguments in defence of the theory that four tastes probably do not account for the whole range of taste sensitivities and that sweet taste itself is a set of various sensitivities. As far as possible, data in the literature are interpreted in terms of receptors rather than neural sensitivities or perceptual qualities; indeed, these are not pure facts but concepts difficult to deal with and, although it may not appear so at first, on a rather highly interpretative level. It must be emphasized that taste quality is a concept which does not correspond with any particular modern electrophysiological or biochemical data.

In Part A we discuss the non-specificity of neural elements and thus the necessity of a spatial pattern as a mechanism for quality coding. Then, the lack of necessity

to find localized projections for qualities will be discussed after considering the multimodal aspect of the sensory image including taste, somesthesia, thermal sensitivity and olfaction. This is a sort of grounding in preparation for the more specialized developments in the second part.
In Part B we discuss the multidimensional aspect of the sensitivity to "sweet taste" which directly challenges the concept of primary tastes. Converging arguments concerning multiple sensitivities to sweetness come from (a) observation of interspecies, interindividual and interfibre differences of sensitivity to selected sweeteners, (b) the differential effect of sweet taste inhibitors (which depend on the subject and the sweetener), and (c) cross-adaption data. Multiple sensitivities of the sweet taste suggest the existence of several receptor sites, an hypothesis which might well fit the very latest biochemical findings.

2 Part A: Physiology of Gustation

Taking for granted the existence of "four basic tastes", biologists and neurobiologists have never ended their search for specific organs or neural elements for each of these taste qualities, at every step along the sensory chain and just as soon as technology made an experiment viable. In other words, they looked for physiological correlates of the four distinct qualities, hypothesizing specific peripheral mechanisms to receive the external information and wired lines to transmit it. (This hypothesis would work as well with more than four qualities.) But we shall see that the mechanism of taste discrimination does not rest on such a simplified basis.

2.1 Non-Specificity of the End Organs (Papillae) and of Each Neural Element Throughout the CNS

2.1.1 Papillae

Papillae have been known to mediate each one of several taste qualities since the minute psychophysics of Kiesow (1898). Unfortunately, Von Bekesy (1966) supported the contrary idea (i.e. specificity of papillae) probably because of an inadequate stimulation. But Harper et al. (1966), Bealer and Smith (1975), Kuznicki (1978), Cardello (1978, 1981) and others (McCutcheon and Saunders 1972; Sandick and Cardello 1981) have proven that every papilla is sensitive to several of the four taste qualities (see Table 1).
All authors remark the great difficulty subjects have in perceiving quality when stimulating only one papilla and hence the very high concentrations necessary. We ourselves have stimulated restricted areas of the tongue and noticed the low intensity perceived as a consequence. "Perhaps", Harper et al. wrote in 1966 (p. 324), "the information contributed by only one papilla is a less adequate mediator of sensations than the combined information from several as it happens in normal tasting behaviour. It would seem that each fungiform papilla contributes some

Table 1. Results of experiments showing sensitivity of papillae to different stimuli

	Number of papillae sensitive to:					
	4 Stimuli	3 Stimuli	2 Stimuli	1 Stimulus	None	Total
Bealer and Smith (1975)	5	5	0	3	2	15
Cardello (1981)	32	7	–	1	–	40

information about most stimuli; but that reliable discrimination among many different stimuli usually requires the simultaneous participation of a number of papillae".

2.1.2 Sensory Cells

Sensory cells are the physiological units which are gathered within papillae. They constitute the most unitary element the neurophysiologist can reach. The next functional element is the chemoreceptor molecule itself.

A number of researchers have recorded intracellular responses of taste sensory cells to chemical stimulation, among them Kimura and Beidler (1961) in the hamster, Ozeki and M. Sato in the rat (1972), West and Bernard in the mudpuppy (1978), and T. Sato and Beidler in the frog (1975) (see the review from T. Sato 1980). They all showed that single cells responded to more than one taste quality.

The results in Table 2 clearly indicate how unspecific the sensory cells are: 65% of responses were obtained for a total of 109 cells tested with four stimuli. One may wonder how many cells would have been sensitive either to all stimuli or to one only if a wider set had been chosen. Cells responding to four stimuli would have had fewer chances to respond to all stimuli and would thus not be considered as totally unspecific. Cells responding to only one stimulus would have had some chance to respond to more than one.

Tonosaki and Funakoshi (1984a) presented the responsiveness profiles of 29 mouse taste cells adapted to distilled water to 0.5 *M* sucrose, 0.1 *M* NaCl, 0.02 *M* quinine HCl, 0.01 *M* HCl. These profiles indicate the multiple sensitivity of taste cells in the mouse: only three cells did not respond to NaCl, two did not respond to quinine-HCl, two to HCl; all cells responded to sucrose.

Table 2. Results of an experiment showing single-cell responsiveness of the rat to different stimuli. (Ozeki and Sato 1972)

Single cells responsive to: 4 Stimuli	3 Stimuli	2 Stimuli	1 Stimulus	Total (*n*)
22	40	28	19	109

Assuming that the response probability to each of the four basic taste stimuli in each gustatory receptor cell is independent of the responses to the other three stimuli, the expected probability of responsiveness to any combination of the four taste stimuli can be calculated as the product of independent probabilities. It appears that the observed number of cells which are responsive to each of these combinations is in agreement with the number expected with independence (Ozeki and Sato 1972).
This means that the different sensitivities are randomly distributed among the cells, i.e. that receptor molecules eliciting each of the four qualities are distinct and randomly distributed on the cells. "How the higher nervous centers discriminate information concerning taste qualities that arise from the complex taste receptors (cells) is not yet clear", wrote Kimura and Beidler (1961, p. 137).

2.1.3 Afferent Nerves

The information generated in the sensory cells is collected by afferent neurons. Slow potential variations are than translated into frequency variations of emitted action potentials. Fibres innervating the anterior two thirds of the tongue, first gathered in a chordolingual nerve, are quickly divided into the lingual nerve, which is a branch of the trigeminal nerve (Vth) and the chorda tympani (C. T.) nerve. The C. T. nerve joins the facial motor nerve (VIIth) before entering the skull and is named VIIth bis. Such a separation between somesthesis (Vth) and gustation (VIIth/bis) characterizes the anterior part of the tongue. The posterior part is innervated by the glossopharyngeal nerve only (IXth) which transmits both modalities.
In the rat, fibres innervating the fungiform papillae on the anterior tongue have been shown, to develop an extensive branching pattern (Miller 1971). Several papillae are innervated by one and the same branched fibre, and one and the same papilla is innervated by branches originating from several fibres. Could the intermingled information originating from non-specific sensory cells be sorted out and categorized by this complex circuitry? Could C. T. fibres emerging from such an entanglement exhibit a better selectivity than sensory cells?
On the contrary, the lack of specificity of afferent neurons is an old established fact, as C. T. unitary recordings have been collected long before intracellular recordings were performed (Zotterman 1935; Pfaffmann 1939). As early as 1941, Pfaffmann showed that one fibre could be activated by several of the four stimuli, and that the qualitative sensitivity determined for each fibre should depend on the set of applied stimuli (see Table 3).
Table 3 shows an overall 63%[1] responsiveness among (17+25=) 42 C. T. fibres: this is similar to what is obtained with sensory cells. Ogawa et al. (1968) found no specific fibre at all out of 50 rat C. T. fibres or 28 hamster C. T. fibres.
As early as 1941, Pfaffmann observed that "different endings possess different thresholds. ... In such a system, sensory quality does not depend simply on "all

[1] All 42 fibres responding to all four standards would give (4×42 =)168 responses. The actual data give 106 responses
((9×4)+(3×3)+(3×2)+(2×1)+(1×4)+(6×3)+(13×2)+(5×1))

Table 3. Results of experiments showing responsiveness of nerve fibres to different stimuli and water

	Number of fibres responsive to:					
	4 Stimuli	3 Stimuli	2 Stimuli	1 Stimulus	Water	Total
Fishman (1957) Chorda tympani	9	3	3	2	–	17
Erickson quoted by Frank and Pfaffmann (1969) Chorda tympani	1	6	13	5	–	25
Frank and Pfaffmann (1969) Glossopharyngeal	2	5	12	8	4	31

Table 4. Predicted and observed responsiveness of nerve fibres of the hamster. (Frank 1973)

Stimulus	Predicted responsiveness	Observed responsiveness
Sucrose alone	3.7	9
NaCl or Q-HCl or HCl	21.2	27

or nothing" activation of some particular fibre groups alone but on the pattern of other fibres active" (p. 255). Although non-specific, the responsiveness is preferential. Since then, authors have comparatively quantified the responses of fibres to each stimulus, as they could not assess an "all or nothing" responsiveness criterion (Frank 1973; Mistretta 1972; Ogawa et al. 1968).

Let us note that a high level of responsiveness implies that the more discriminating part of the information is to be found in the lack of response of some cells rather than in the increase of spontaneous activity of nearly all cells, whatever the stimulus might be. It is well possible that the basis of the signal should consist of the lower level of excitation of some cells to a given stimulus in contrast to other cells.

Frank's results on the hamster (1973) agree with Table 3. However, Frank noticed a difference between the observed numbers of fibres responding to sucrose alone and to the group of stimuli Q-HCl, HCl, NaCl, with respect to the numbers predicted by the theory of independence of all four stimuli (Table 4).

The deviation from stochastic independence indicates some overlap of sensitivity, some communal acceptor mechanism for both Q-HCl and HCl – which is obvious since the concentrations of each of the stimuli were similar (0.001 *M* and 0.003 *M*, respectively) – and, NaCl as well. Ionic sensitivities probably do not use chemoreceptor mechanisms but ions may directly interact with the cell ionophores; Na^+ and H^+ may partly share some common excitation mechanisms, independent of the receptor mechanisms involved in the sensitivity to organic molecules. The tendencious choice of the four stimuli: HCl, Q-HCl, NaCl which involve ionic stimulation, vs sucrose which is purely organic, may explain the differences observed by Frank and other authors.

Ninomiya et al. (1984) recorded mouse C. T. fibres and apparently found groups of narrowly tuned fibres in this animal. Upon first reading the paper might be considered as an example of a counter-argument to the theory of non-specificity of taste fibres. But, once again, a bipolarity is obvious, with fibres either responding to sugars (sucrose, glucose, fructose, maltose) and saccharin, and not to ions (Na^+, K^+, NH_4^+, Ca^{++}, Mg^{++}, H^+ or quinine-HCl), or vice versa. Quinine-HCl-responsive fibres, respond to Ca^{++}, Mg^{++} and acetic acid.

This result strongly confirms the hypothesis of two families of mechanisms. One of these involves proteic chemoreceptors for the recognition of a huge number of organic molecules – sweet or bitter tastants, or others (which are not screened when only five stimuli among which four are "sugars" are used) – and the other a mechanism for pH and ionic stimuli. In electrophysiological experiments Q-HCl may not conveniently represent the bitter taste as perceived by humans.

Ions potentially acting on the general physiology of sensory cells are probably not appropriate for the study of chemoreception.

Random Sensitivity. Some discrepancies are found in the literature concerning the randomness of the sensitivity distribution among fibres, i.e. the independence of the four tested sensitivities among fibres. Frank and Pfaffmann (1969) showed data supporting a random sensitivity to four taste qualities in the glosso-pharyngeal nerve as well as in the C. T. of the rat. Ogawa et al. (1968) found a correlation between quinine and HCl on 50 rat C. T. fibres which could probably not be extracted from the 25 C. T. fibres Frank and Pfaffmann used in their calculations. This correlation is however questionable as the bitter stimulus was not quinine but quinine hydrochloride (of comparable molarity to the HCl solution they used as an acid stimulus). In the hamster, Ogawa et al. (1968) found an anti-correlation between sucrose and NaCl which, although weak (−0.4), is significant.

Frank (1973) failed to confirm this negative correlation between NaCl and sucrose on 79 hamster C. T. fibres. It might well be that the Ogawa et al. fibre sample was not representative of the whole nerve; adding more unspecific fibres could add variance and downgrade the correlations. A tentative conclusion might be that there is independence of neuron sensitivities to the four standards, hence, independence of their peripheral chemoreceptor mechanisms.

2.1.4 Geniculate Ganglion

The geniculate ganglion gathers the cell bodies of the C. T. fibres and can be conveniently accessed through a surgical opening of the skull and petrous bone. Boudreau and co-workers extensively recorded in this way, using a wide variety of stimuli. Boudreau and Alev (1973) and Boudreau et al. (1975) tested 141 qualitatively different aqueous solutions on geniculate ganglion neurons in the cat. Not unexpectedly, they found much difficulty sorting out taste units according to their chemical sensitivity, as each unit not only had its own spectrum of sensitivity but also often failed to consistently respond to a repeated stimulation.

Their attempt to classify units was better achieved when they used amino acids, nucleotides etc., instead of the four traditional reference stimuli.
Boudreau usually classified units with respect to electrophysiological and phenomenological properties, such as latencies, bursting activity etc. He did so recently for the goat, finding no indication of a grouping of fibres with respect to any chemical specificity nor to any quality specificity (Boudreau et al. 1982).

2.1.5 Nucleus Tractus Solitarius

The nucleus tractus solitarius (NTS) is the first relay in the taste afferent pathway. Units recorded in such a structure are not easily identified; they may be presynaptic or postsynaptic afferents or even any kind of interneuron devoted to intranuclear data processing. All units recorded in the nervous centres we shall deal with are subject to this ambiguity.
Travers and Smith (1979) found hamster NTS units even more broadly tuned than C. T. fibres are as far as their sensitivities to the four reference stimuli are concerned. Doetsch and Erickson (1970) had previously shown better discrimination in the NTS than in the C. T. between stimuli which elicit qualitatively different perceptions in the human, and with respect to stimuli eliciting similar perceptions. Such an improved discrimination must obviously rely on other mechanisms than the selectivity of single neural elements.

2.1.6 Pontine Taste Area

The pontine taste area (PTA) which was discovered by Norgren (1971), Norgren and Leonard (1971a, b) and Bernard and Nord (1971) had been previously identified in the carp by Herrick in 1905.
This area, which has been examined anatomically by a number of authors (Norgren 1976, 1978; Lasiter and Glanzman 1983; Lasiter et al. 1982; Nomura et al. 1979) presents no decussation (Goto et al. 1983) – contrary to descriptions in some textbooks – which explains the unique homolateral projection of taste-sensitive fibres in the thalamus (Benjamin 1963).
The PTA offers the same disappointment to those expecting to find some increased specificity of single neural elements to account for the quality discrimination process. Although there is a possibility to group neurons with respect to a higher response to a preferential stimulus among four (Smith et al. 1983a; Ogawa et al. 1984), the tuning of the cells is even broader here than at the NTS levels (Smith et al. 1979; van Buskirk and Smith 1981); see Table 5.

Table 5. Results of an experiment showing responsiveness of pontine taste area neurons to different stimuli. (Scott and Perrotto 1980)

Number of PTA neurons responsive to:				
4 Stimuli	3 Stimuli	2 Stimuli	1 Stimulus	Total
21	10	7	2	40

That the cell tuning may be broader does not mean that interquality discrimination is not already accomplished. Perrotto and Scott (1976) have shown that stimuli which taste alike to man are correlated across rat PTA single neuron responses; moreover, they have shown that profiles of neuron responses to various concentrations of the same stimulus are highly correlated (Scott and Perrotto 1980).

Hence, quality coding is probably not accomplished by classification into four classes of responses, but by another process (which will be developed in Section 2). Schwartzbaum and Block (1981) confirm the non-specificity of PTA single units. They have reported that "even the most narrowly tuned cells, those designated as NaCl dominant and sucrose dominant, could not *per se* distinguish by their magnitude of response whether the stimulus was a high concentration of some alternative tastant or a low concentration of the otherwise most effective tastant. The pattern of response across elements had to be utilized." Smith et al. (1983b) also agree with this opinion. According to them, "particular sets of neurons are critical for coding the tastes of particular classes of stimuli, but activity in more than one group of neurons is necessary for the unambiguous coding of taste quality."

Thus, gustatory information is apparently similarly processed at all points from the chorda tympani to the PTA.

2.1.7 Thalamus

Authors such as Norgren (1977) and Yamamoto et al. (1980b) clearly described a Y departure for two gustatory pathways from the PTA in the rat: first, a projection to the cortex via the thalamus (Benjamin and Akert 1959) and second, a projection to the hypothalamus, amygdala and bed nucleus of the stria terminalis. But, the thalamus is sometimes by-passed. Lasiter et al. (1982) showed the existence of a direct projection from the PTA to the cortical taste area which parallels the first pathway.

Benjamin (1963) and Norgren and Wolf (1975) localized the thalamic taste area in the more medial, ventral and posterior part of the homolateral ventro-postero median nucleus, i.e. the so-called VPMpc. It must be noted that the projections of the C. T. and IXth nerves are bilateral as far as mechanical or thermal sensitivities are concerned, but homolateral for the purely chemical sensitivity.

Single thalamic neuron recordings lead to the idea that the coding process might not rest on the same basic rules in the thalamus as in the brain stem (NTS, PTA). Researchers report poor and sluggish responses to the four basic taste stimuli in the thalamus. Responses to quinine hydrochloride and sucrose are as important as are responses to sodium chloride and hydrochloric acid, which was not the case in the lower parts of the sensory chain. They find that the responses to stimuli within a group of the same quality are less closely correlated than in the brain stem or in the C. T., whereas the stimuli representing different qualities are not so strictly separated from one another as if fine discrimination within one quality was better processed in the thalamus than in the brain stem (Scott and Erickson 1971). Scott and Yalowitz (1978) suggested that some thalamic cells could be specific not only to one quality (i.e. one stimulus among the four tested ones), but

Table 6. Results of an experiment showing responsiveness of thalamic neurons of the dog to different stimuli. (Ninomiya and Funakoshi 1982a)

Number of neurons responsive to: 4 Stimuli	3 Stimuli	2 Stimuli	1 Stimulus	*n*
8	17	29	14	68

to a given intensity as well. In the afferent nerves, as in the brain stem, both intensity and quality are encoded within every cell; the firing rate of each cell is altered by a change in quality as well as in concentration of the stimulus. Conversely, Scott and Yalowitz found five out of 23 thalamic neurons to be sensitive to one or two qualities whatever the concentration, and nine other neurons which, although sensitive to several of the four stimuli, exhibited an increasing intensity response function for only one stimulus. Five other neurons were non-specific, as lower order neurons are, and four had no consistent responses. In the dog, Funakoshi and Ninomiya (1977) and Ninomiya and Funakoshi (1982a) found about the same: only 20% among 59 and 68 neurons, respectively, were specific for one stimulus among four (Table 6).

2.1.8 Cerebral Cortex

As early as 1957, Landgren had reported that cortical taste cells exhibited a multiqualitative responsiveness to taste stimuli. However, Funakoshi et al. (1972) found a majority of cells responding to only two or even one of the four stimuli (see Table 7).[2]
Among other technical reasons, the differences between the several author's results may arise from different response criteria as shown in Table 7 by data from

Table 7. Results of experiments showing cortical cells responsiveness to different stimuli

	Number of cells responsive to:				
	4 Stimuli	3 Stimuli	2 Stimuli	1 Stimulus	*n*
Funakoshi et al. (1972)					
– By excitation	2	3	10	18	33
– By inhibition or by both excitation and inhibition	4	1	2	2	9
Total	6	4	12	20	42
Yamamoto et al. 1981d, e					
– 1 s.d.	16	16	8	8	48
– 3 s.d.	2	2	11	18	33

[2] See Braun et al. for a review on the gustatory neocortex (1982)

Yamamoto et al. (1981e). These authors found 67% of units responsive to three or four stimuli using a criterion of 1 s.d. above the mean, whereas 12% only of the same cells respond to three or four stimuli using a criterion of 3 s.d. above the mean. The same comparison at the C. T. level does not lead to such a difference; the authors suggest that cortical neurons could actually be less broadly tuned than C. T. neurons.

Ninomiya and Funakoshi (1982b) found cortex cells less broadly tuned than thalamus cells in the dog, but, the preparations differed; cortical cells were recorded in awake dogs and thalamus cells in anesthetized animals. The authors further showed that the parameter evaluating the breadth of responsiveness of neurons was sensitive to normalization of responses to the level of spontaneous activity which was actually different in the two structures (Funakoshi and Ninomiya 1983).

Yamamoto (1983) and Yamamoto et al. (1984c) found no significant difference in the rat between the breadth of responsiveness of 106 cortical neurons and 50 C. T. fibres. Yamamoto et al. (1984c) had previously reached "an opposite conclusion on the basis of an all-or nothing measure using an arbitrary response criterion for an occurrence of response". They attribute their earlier conclusion to "two properties of cortical taste neurons: their evoked rates are lower and their spontaneous rates higher than the chorda tympani fibres".

In accordance with the calculations of Scott and Erickson (1971), Yamamoto et al. (1981e) found that the response profiles to stimuli of different taste qualities exhibit much higher correlation coefficients than profiles obtained at the C. T. level. This could mean that the four reference stimuli are relatively less discriminated in the thalamus and in the cortex than at lower levels. However, Yamamoto and Kawamura (1977) had also previously shown that the four sensitivities were randomly distributed among cortical cells as in other parts of the sensory chain.

Conversely, it seems that quality discrimination is more powerful at higher levels than at the brain stem level: "the small qualitative differences are analysed as well as the larger ones" which were already discriminated in the brain stem.

In these conditions, the quantitative correlation coefficient is not a good measurement of quality discrimination which does not appear any more as a quantifiable distance. If qualities are all distinguished at cortical level, stimuli profiles across units are all independent and the *r* value is no indication of the similarity between stimuli, as far as their activity on the receptors is concerned. This situation is different from the measurements of perceived qualitative distances (for example: Schiffman et al., 1979) were the *r* value is actually an adequate quantifier of the paired similarities.

Summing up these data, we find some difficulty in comparing the results obtained from different authors because of the use of different response criteria. Some data are qualitative, i.e. if the level of excitation overcame a statistical criterion, the cell was said to respond to the presented stimulus. Others are quantitative, i.e. the authors counted the number of elicited spikes during stimulation, irrespective of any statistical criterion, based on the varying spontaneous activity. Indeed, it is difficult to know which is better. Even if the activity following application of the stimulus does not overcome the predetermined criterion, we can never be sure that a small excitation, a few supplementary spikes, simultaneously occurring in a number of fibres, is not taken into account by the CNS in the coding process.

The distribution of sensitivities to the four standards seems random at any level of the sensory chain. This would mean that these four sensitivities arise from independent receptor mechanisms at the interface with the external medium. From the C. T. level to the pontine taste area level, the breadth of tuning of single units increases, hence, quality cannot be simply encoded in specific groups of fibres or cells. Thalamic and cortical taste areas possess some cells which seem to be more specific – we could say less unspecific since no more than four stimuli have ever been tested – than cells in lower parts of the sensory chain. Nonetheless, it should be emphasized that only 55% of cells are found to be specific using the most favourable criterion (e.g. Yamamoto et al. use 3 s.d.). This explains how the response profiles to different qualities can be even less distinguished in the cortex than at lower levels, although more specific cells can be found in the cortex. Whether the higher centres will take care of the whole message, or only of the information coming from a few outstandingly specific cells, cannot be guessed. Several authors, among them Maes and Erickson (1984), take the position that the coding including only more specific fibres (the so-called sucrose-best-responding fibres, for example) is not discriminatory and that we must use all fibres to model a population of responses able to distinguish between stimuli of distinct qualities. And last but not least, nobody has ever tested the extent to which the percentage of central specific units could be reduced by using more than four stimuli.

2.1.9 Problems Elicited by Neuron and Stimulus Classification when More than Four Stimuli Are Tested

However non-specific, neurons often exhibit a preferential sensitivity towards one of the four standards: sucrose, NaCl, HCl or quinine-HCl; hence, it is worth trying to classify neurons under such labels as sucrose-best, HCl-best, NaCl-best or quinine-HCl-best, as first suggested by Frank (1973). We shall see that it is a convenient mode for representing the data if the four stimuli only have been used in the experiment, i.e. when there are no more stimuli than labels. One can wonder whether a four-termed classification is justified and whether the four usual stimuli are good representatives of these four categories. In fact, difficulties arise as soon as the stimulus set becomes widespread.

Frank (1973) calculated the correlation coefficients between response profiles to the four standards and other stimuli across 46 C.T. fibres (p. 605). Let us successively examine the cases with saccharin, KCl and $MgSO_4$.

Saccharin appears to be correlated to sucrose only, and the coefficient is not high (0.61). Although it is possible to compare saccharin sensitivity to sucrose sensitivity, saccharin is not equivalent to sucrose for the whole set of fibres; hence it is not completely described by the classification using these four labels. KCl is not correlated with sucrose (−0.14), but is equally (loosely) correlated with NaCl (0.52), HCl (0.59), quinine-HCl (0.58). There is no way of classifying KCl by reference to the four basic tastes except that it is fully independent of sucrose and equally loosely correlated to the other three stimuli, i.e. many sucrose-sensitive fibres may or may not be responsive to KCl with a full random probability, whereas fibres which are sensitive to any of the other three stimuli will equiprobably respond to KCl. $MgSO_4$ also is not highly correlated to any of the four stimuli (HCl, $r = 0.57$, QHCl, 0.48, NaCl, 0.39, sucrose, −0.03). The higher correlation is found with HCl. Although significant, this correlation is loose. Hence, $MgSO_4$ is distinct from any of the four stimuli for any fibre. The low coefficients demonstrate a certain independence of $MgSO_4$ towards any of the four labels which are used here. However, it is interesting to consider the quantitative variation of these correlation coefficients. Indeed, the correlations rank from 0.57 to −0.03 from HCl to sucrose, here the usefulness of a continuous parameter for measuring similarity of responses of single units to several stimuli appears. This is just the contrary of the idea of a classification!

Could a higher correlation be found between $MgSO_4$ and let us say L-glutamate, if Frank had added, as a fifth label, this representative of the "umami" taste – which is known in Japan to be independent of the four basic tastes (Yoshida and Saito 1969; Yamaguchi 1979)? Is it of any use to classify fibres with respect to four categories, if this classification is not predictive of their sensitivity to other stimuli? How many labels must we use to be able to classify taste fibre responsiveness with respect to a great variety of stimuli? Is it possible to find a classification at all, or will we have to use as many labels as stimuli? In other words, are the spectra of C.T. fibre responses all different, or can we obtain any grouping at all of similar fibres?

The controversy is presently open and needs, perhaps, a much wider sample of fibres stimulated by a much wider set of stimuli than has ever yet been gathered. In a review, Pfaffmann et al. (1979) quotes C.T. data from Erickson et al. (1965) in which he can find "three clusters although Erickson has continued to maintain that obvious receptor types of clusters are not apparent" (p. 287). Let us remark here that a rectangular matrix of fibre responses to various stimuli can be analysed using paired correlations either across columns (correlations between paired stimuli across all fibre responses) or across lines (correlations between paired fibres across their responses to all stimuli). The first answers to the question "How similar are pairs of stimuli?", the second to "How similar are pairs of fibres?". Erickson et al. (1980), using hierarchical clustering as well as both kinds of paired correlations, found groups of stimuli, just as Pfaffmann did on the same kind of data set, but no groups of fibres. Pfaffmann calculated paired correlations only between the stimuli, not between the fibres, and hence cannot state having found fibre grouping.

Although a classification either of fibres or of stimuli within the four quality labels is never satisfactory, it is eventually possible to find some grouping of stimuli depending on their nature. In the data of Erickson et al., Na and Li salts are correlated and thus form a group, while KCl, NH_4Cl, $MgCl_2$ form another one. As a matter of fact, the other stimuli do not form any group since they are all more or less independent. (One can see this in the data of Pfaffmann et al. 1979.)

Chemicals can be chosen by the experimenter on the basis of their physicochemical similarity or of their qualitative similarity as perceived by the human. Hence, the profile of activation of animal fibres may be found similar. We can more easily extract similar stimulus profiles than similar fibre profiles from neural data because, in the experimental procedure, we can choose the stimuli, whereas we can never choose the units!

2.1.10 Different Stimulus Profiles and Implications at Receptor Level

If all sensitivities were randomly distributed among neurons (and this seems to be the case throughout the taste nervous system, at least for the four standards), a great variety of sensitivity spectra could be found among neurons. How many? This strictly depends on the number of independent peripheral receptor mechanisms; we should find as many different sensitivity spectra as combinations of independent receptor mechanisms (2^n *if* n is the number of receptor mechanisms). This number remains presently unknown and we shall see the beginnings of a response to this question in Part B.

In a practical sense, the number of different sensitivity spectra depends on the size of the data matrix, that is on the number of stimuli and fibres. Using twenty-five units stimulated with 12 stimuli (Erickson et al. 1980) does not reach the probability to find similar profiles, whereas 50 units stimulated with only four stimuli (Ogawa et al. 1968) does.

All kinds of sensitivity profiles to 12 stimuli are not sampled by 25 units, whereas all kinds of possible profiles to four stimuli are sampled by 50 units. Yet, neither four nor 12 stimuli sample all actual sensitivity profiles.

Finding similar across-fibre profiles for different stimuli implies that peripheral acceptor mechanisms are not strictly specific and that there is not an independent receptor to signal each chemical.

2.2 Pattern Theory

2.2.1 Quality Coding

If there is no group of specific neurons, another coding process must exist.

Pfaffmann (1939, 1941, 1955, 1959) was the first to realize that neurons could code sensory quality through a pattern of excitation of varied intensity across *all* neurons. As early as 1939, Pfaffmann understood from his own data that "fibre categories are based, of course, upon four types of stimuli used for testing. There is evidence that certain other substances may also stimulate more than one fibre so that if a wide variety of agents were used, each fibre might be found to have a chemical "spectrum" which overlapped those of other fibres. In 1955, he suggested that "in taste, the afferent neural message for quality is probably expressed in terms of the relative amount of neural activity across neurons." Erickson took to this subject and brought conceptual as well as experimental arguments to what he called the "across-neuron pattern" theory. In 1963, Erickson had already concluded from his own experimental work on taste that "they are many fiber types representing gustatory quality as for pitch discrimination, rather than a few fiber types as in color vision". Since then he has been defending this theory with psychophysical as well as electrophysiological experiments (Erickson 1968; Erickson et al. 1965; Doetsch and Erickson 1970; Schiffman and Erickson 1971;

Woolston and Erickson 1979; Erickson and Covey 1980; Erickson 1982, 1984). Other authors, too, have brought some comprehensive arguments to this issue, including Maes (1984), Maes and Erickson (1984), Yamamoto and Kawamura (1972), Schwartzbaum and Block (1981), DiLorenzo and Schwartzbaum (1982a), and Smith et al. (1983a,b). A rather wide population of researchers now take it for granted that taste coding relies on an across-neuron pattern of activation, and use this theory as a starting point in their own data interpretation.

DiLorenzo and Schwartzbaum (1982a) showed that if there are a few narrowly tuned neurons in the pontine taste area, "these neurons alone could not effectively code specific taste qualities".

Smith et al. (1983b) in a paper on the hamster brain stem neurons wrote: "Although a given neuron group is responsible for the similarities in the patterns elicited by a particular group of stimuli, no one class of neurons is sufficient to establish different patterns between dissimilar-tasting compounds. Thus, the neural distinction between many behaviorally discriminable stimuli depends on the simultaneous activity in different classes of neurons".

The pattern theory states that there is no quality coding in an unique fibre, but in a population of them. The CNS extracts the quality message from the relative activity among all afferent fibres through a process of form recognition (spatial pattern).

Imagine you are faced with a screen dotted with as many lamps as there are fibres in the chorda tympani nerve. As a first hypothesis consider that each lamp can be lit or not depending on whether the fibre it represents is responsive to the applied stimulus or not. The global result is a pattern of lit and unlit spots. We can imagine that in this way the quality message is transmitted to the CNS. For each different stimulus, the spatial pattern is different although a great many activated fibres may be common to several patterns, e.g. although many fibres respond as well to NaCl as to sucrose, the global pattern will be different and signal two different qualities.

As a second hypothesis, consider that the lamps do not represent an all-or-nothing function, but emit spots of variable intensity. Hence, the difference beteen two patterns emerges from the relative intensity of the spots, which are nearly all lit in both cases. The second hypothesis fits best with data in the literature on single-unit recordings and is able to account for the spontaneous activity of neural elements as well as for inhibitory responses which do exist even in the C.T. (Scott and Pritchard 1982).

2.2.2. *Reproducibility of the Pattern*

Since nearly no fibre is specific and because quality information results from a relative activation among fibres, it is of some interest to question whether the constancy of the message elicited by replicated stimulations rests on a high reproducibility of the unitary responses. In the hypothesis of a classification into best-responding fibres, a sucrose-best-responding fibre should not respond best to NaCl – or to sucrose from time to time.

Ogawa et al. (1973) showed that the reproducibility of C. T. single-unit responses is not good. Boudreau et al. (1975) failed to find good reproducibility in the geniculate ganglion, and Van Buskirk and Smith (1981) showed that PTA neurons could change class "with concentration level and sometimes when the four stimuli were repeated". Hence, we must deal with patterns the stability of which does not rest on the constancy of each neural element. Conversely, the pattern relies on the high number of its constituting elements so that each unit is allowed some variance. In other words, one and the same pattern is not always constituted of the same neurons exactly; although there are uncertainties and blurred outlines, the pattern is still recognizable. This mode of functioning which extracts signal from noise powerfully allows analogies.

2.2.3 *Intensity Coding*

In this theory, intensity is coded both through the higher excitation of each neural element and the excitation of a higher number of neural elements (Yamamoto and Kawamura 1972; Maes 1984; Maes and Erickson 1984). Hence, less sensitive units, with perhaps broader response spectra, are recruited when concentration increases. We understand that units which have been shown to be less specific could be responsible, above all, for intensity coding (Yamamoto and Kawamura 1972).

2.2.4 *Stability of Quality with Change in Intensity*

The recruitment of less sensitive units possessing a broader sensitivity spectrum can alter the qualitative pattern. This has been shown by electrophysiologists (Yamamoto and Kawamura 1972; Ogawa et al. 1974; Ganchrow and Erickson 1970) and is well known from psychophysics, too. The saccharin taste changes with concentration; NaCl is nearly sweet at threshold level, etc. As far as NaCl, HCl, quinine and sucrose are concerned, and within some concentration limits, the qualitative distances between several concentrations of the same stimulus are always smaller than distances between different stimuli (Scott and Perrotto 1980). The adequacy of blurred patterns for quality coding appears here.
A very large number of different patterns, limited only by readibility of activity level in a given neuron, can thus be coded without single neuron specificity but just preferential sensitivity, without constancy of the unitary signal. Let us compare electrophysiological unitary recording with human judgments obtained in psychophysics. Quantitative assessments are never achieved with only one response from only one judge. Conversely, if the test is repeated n times, the expected value measuring the perceived intensity is known within some limits. The great number of neural elements, functioning all at the same time and constituting a pattern of activated/non-activated cells work in the same way as n repetitions of the same response, from a statistical point of view. As variance is in inverse ratio to $\sqrt{n}$ (n is the number of neurons), the more elements there are, the less the variance. Responses to NaCl are reproducible within 10% in integrated C. T. (Beidler 1953). A hundred fibres each presenting 100% variance could account for this global variance (100% divided by $\sqrt{100}$ is 10%.) This means that if a subject has only 100 sucrose-responding fibres, and if each of these fibres responds with a variance of 100% the subject will still properly recognize sucrose.

As man possesses about 2000 fibres per chorda tympani, there is good assurance even with non-constant unitary responses.

2.2.5 Interquality Interactions: Mixtures

If many different combinations of activated cells can be constituted within the same set of neurons, one can wonder what happens if a mixture, not a pure stimulus is applied. The pattern resulting from the addition of two stimuli is not the result of the mere addition of the patterns that each stimulus would have elicited. Because most of the neurons are not specific, there is a high percentage of overlap, and adding two stimuli gives rise to a third, different, pattern. The union of two sets might even not concern many more neurons than each of the sets independently. This question is being electrophysiologically addressed (Hyman and Frank 1980a, b; Travers and Smith 1984).

The psychophysical correlate of this electrophysiological phenomenon should lie in a synthetic continuum of taste sensitivities. In a synthetic modality like colour vision, the perception resulting from a mixture of two components is a synthetic perception differing from the perception elicited by each of the components: one cannot recognize yellow and red in an orange colour obtained by mixing two appropriate wavelengths. In an analytic modality, on the contrary, the subject can easily recognize the constituents of the mixture. It is so with audition; two different notes are identifiable in a chord. Erickson and Covey (1980) asked subjects to measure the "singularity" of their perceptions while tasting either pure solutions of sucrose, salt, etc. or mixtures of them. The result is surprising, since a pure chemical is never perceived absolutely as "singular". A percentage varying from 10% to 50% of responses indicated that subjects perceived more than one taste in a solution of a pure chemical; a mixture such as $MgCl_2$-KCl or Q-SO_4-HCl was perceived 50% of times as "singular", just as the pure components are.

Hence gustation appears to be neither totally synthetic nor analytic. Indeed, it is possible to recognize mixture components after a training to get acquainted with them, but the mixture does not taste just like the simultaneous perception of each component. Depending on the focus of our attention, we are able to consider either the patterns elicited by components or the global pattern elicited by the mixture.

We can focus either on the similarities or on the dissimilarities of a given pair of stimuli, should they be pure chemicals or mixtures, on request. Is thaumatin sweet? Yes, of course. Let's ask another question. Is thaumatin sweet like sucrose? The response is often no, depending on the subject and the concentration. The same is true for a mixture. Any new composition is a new form to the CNS, which is unique. But the CNS can analyse it by analogy in terms of partly similar, partly dissimilar to one or two of its components (Bartoshuk 1975; O'Mahony and Buteau 1982).

Psychophysicists are often questioned about the intensity resulting from mixing stimuli; this appears to be a major problem in industrial applications. The search for sensory synergies is very up to date. So far, only monosodium glutamate seems to be synergized by nucleotides (Yamaguchi 1967, 1979); this is confirmed by

Rifkin and Bartoshuk (1980); while an exhaustive review (of mainly Japanese papers) is provided by Maga (1983).
Some mixtures result in an hedonic improvement, e.g. salt plus monosodium glutamate is not more salty but only more palatable (Yamaguchi and Takahashi 1984) than salt alone.
A series of models to try to predict the intensity of a mixture have been published (Bartoshuk 1975; Bartoshuk and Cleveland 1977; Frijters and Oude Ophuis 1983; Frijters et al. 1984; Curtis et al. 1984). The intensity obtained when mixing stimuli is usually below algebraic additivity; opposite results are measurement artefacts, except for monosodium glutamate and nucleotides (McBride 1984).

2.2.6 Temporal Aspects of Electrophysiological Response and Behavioural Correlates

Unitary and averaged electrophysiological recordings exhibit a temporal change of the firing pattern. Is this temporal pattern of any use in coding chemical information? Should we thus consider a spatiotemporal pattern for quality coding? According to Faull and Halpern (1972), chorda tympani response to NaCl consists, after a brief latency (30 ms), in a rapid rise (phasic component) and a slow decline to a sustained level (tonic component).
Behavioural experiments from Halpern and Tapper (1971) have shown that in a conditioned aversion paradigm the rat reaction time to recognize NaCl is less than 600 ms at the first presentation and less than 250 ms at the sixth presentation; this is about the time required to reach the maximum of the phasic part of the C.T. response. This would mean that the following tonic part of the response is not necessary to recognize salt quality. Yamamoto and Kawamura (1984) and Yamamoto et al. (1983, 1984b) measured reaction time of human subjects for different stimuli, 600 ms are sufficient to signal the change of taste for salts such as 0.1 *M* NaCl, 0.5 *M* NH_4Cl etc. If one agrees with Hara (1955) that 2 s are sufficient to recognize each of the four qualities, then the very beginnng of the electrophysiological response is sufficient to recognize quality. When dealing with more complex stimuli, such as thaumatin, however, we can notice that for example 8 s are necessary to reach the maximum response in the monkey C.T. (Hellekant 1975). This is well correlated with human perceptions, where about 10 s are necessary to fully perceive the sweet taste of thaumatin (Birch et al. 1980), whereas the sweet perception elicited by $CHCl_3$ vanishes within 3 s.
The maximum of the electrophysiological response is obtained within 500 ms for NaCl and 500–1000 ms for sucrose (Hayashi 1976). Fishman (1957) and Yamashita and Sato (1965) also noted the characteristic differences in the time course of the integrated responses to various gustatory stimuli in the rat C.T. Ichioka (1972) showed that the time courses of summated C.T. responses are significantly different for each of the four standards. This led Ichioka and Hayashi (1974) to conclude that "neural correlates of taste qualities must be searched for in spatio (across-neuron response) – temporal (early stages) patterns of afferent impulse trains".
A number of authors have observed a periodic pattern in some unitary responses to sucrose (Fishman 1957; Doetsch and Erickson 1970; Mistretta 1972; Frank and Hettinger 1984). The periodicity is about 500 ms, but varies from one trial to

another, and not all sucrose-responding fibres exhibit such periodicity (Ogawa et al. 1973).

Nagai and Ueda (1981) find such rhythmicity in 7 out of 19 sucrose-responding neurons, but it occurs for representatives of other qualities as well. Hence, the temporal pattern seems to depend on the fibre as well as on the stimulus. For Ogawa et al. (1973) there is no conclusive evidence for a temporal aspect of the taste quality coding.

Nagai and Ueda (1981) studied stochastic properties of gustatory impulse discharges in rat C. T. single fibres. They showed that, although a strict classification of the tonic activity cannot be made, bimodal or skewed distributions of interspike intervals (distributions exhibiting regular or irregular bursts of spikes) are more often associated with sucrose, whereas gamma or exponential distributions are more often associated with NaCl. Averaging single neuron responses, they found a typical temporal pattern for the initial dynamic phase for NaCl only, but not for the other three qualities.

A multi-unit analysis of the rostral NTS and parabrachial nuclei activity (Schwartzbaum and DiLorenzo 1982; DiLorenzo and Schwartzbaum 1982b) reveals that the "time course of the response could contribute to neural differentiation of the tastants". The maximum of the response is obtained within about 1 s for NaCl, within 2–3 s for sucrose, HCl or Q-HCl, and the patterns of response magnitudes over time are different for NaCl, sucrose and HCl.

A temporal pattern is not necessary for quality coding, but whether there is no coding within the temporal aspect of the early stages of the response is not proven. Nonetheless, Doetsch and Erickson (1970) have shown that the stability of quality coding over time is obtained after the first synapse. Using only the early stages (0.3 s) of the response, or a longer period, is equivalent when dealing with NTS neuron responses, whereas it leads to very different results in the C. T.

From the current literature it appears that there is no more fluctuation with time in the electrophysiological response beyond the first relay. Finally, no stochastic studies have yet been conducted in the higher centres; thus, we do not know of any temporal patterning there.

As far as the peripheral level is concerned, a putative temporal pattern should lie within the first 300 ms and characterize the phasic component of the response only. What can be the usefulness of a long-lasting excitation in response to chemical stimulation?

First, it is known from psychophysics or behavioural experiments (Du Villard et al. 1981) that intensity needs a longer time to be estimated than quality to be recognized. Moreover, complex stimuli, the quality of which is not recognized at once, or more subtle discrimination, may need a longer time to be identified. If the CNS acts as a spatial integrator, it may integrate information over time and thus provide a better discrimination in a more difficult situation. The time to reach a behavioural response, which may coincide with the time to reach the maximum of the electrophysiological response, must not be considered as a mere latency, but as the necessary time to sum up enough information after which the message may fall into redundancy.

Second, if this long-lasting electrophysiological excitation is of no more use when quality and intensity have been recognized and estimated, it has nonetheless a consequence which is called the adaption phenomenon.

2.2.7 Adaptation

The adaptation of the electrophysiological response consists of the declining course of the response with time after the initial maximum resonse. According to Smith et al. (1975), this time course can be described by two exponential functions which hypothetically correspond to the desorption of the stimulus molecules on the receptor sites, and an intrinsic receptor cell adaptive mechanism. (In fact, it is very difficult to know whether the stimulus has been actually rinsed or not and whether the excitation goes on after the stimulus has disappeared or not. We can only know whether the equation describing the excitation decline is concentration-dependent or not.) By extension, the adaptation is, as well, the reduction of the activation of a neuron to a stimulus applied just after a first one, by comparison to the activation elicited by the same stimulus alone.

By testing salt stimulation in the frog, Kashiwagura et al. (1976) showed that phasic responses are subject to adaptation, whereas tonic responses are not; i.e. only the phasic part of the response to salt after adaptation with salt is lacking. Smith and Frank (1972) and Smith and Bealer (1976) observed the same phenomenon on the rat. Smith et al. (1978) studied the time course of adaptation of both phasic and tonic parts of the rat C. T. response with longer exposure to the adaptant stimulus.

In the frog T. Sato observed an initial phasic depolarization of the receptor potential of taste cells only after adaptation to water (Sato T. 1977), whereas he could never see any phasic response after adaptation to Ringer's solution (Sato T. 1971).

Cross-adaptation is observed, too. Andersen et al. (1963) found a reduced response of single units to sucrose stimulation after NaCl in the dog. Miyake et al. (1976) observed the same phenomenon and Ozeki and Sato (1972) observed it, too, in rat intracellular recordings.

Whether adaptation depends on receptor desensitization or on sensory cell physiological properties is debatable (Smith et al. 1975, 1978). Probably both mechanisms are involved.

The psychophysical adaptation involves a third hypothesis, i.e. there is a central process resulting from the non-specificity of neural elements, where patterns to different stimuli share some common parts, some common neurons. A neuron which is still firing as a result of the previous stimulation will contribute a very poor signal with respect to its previous activity, now considered as noise.

If adaptation can be an object of study (Von Bekesy 1965, Bartoshuk et al. 1964, McBurney et al. 1972, McBurney and Bartoshuk 1973, Kroeze 1979, 1982a,b, 1983, Gillan 1984), then it is also an useful tool to study the receptor mechanisms which are common to several stimuli, irrespective of a full comprehension of the phenomenon (Schiffman et al. 1981, Murphy et al. 1981, Lawless and Stevens 1983).

2.3 Chemotopy

While looking for specificity of taste organs, researchers have also looked for a topical organization of taste qualities (chemotopy) at every level of the sensory chain. It seems that discovering spatial patterning for quality encoding enhanced the hypothesis of chemotopic projections.

Since there are no specific cells, one cannot expect any precise topography, but perhaps a tendency to topographical organization of cells preferentially responsive to a given stimulus. I will summarize the literature which has been written on various parts of the taste system in order to make out a spatial organization of quality. The first thing that will become obvious is that a topical projection of the two mammalian taste nerves (VII bis and IX) is found everywhere in the taste nervous system. As the qualitative sensitivities of the VIIth and the IXth nerves are somewhat different, we shall see that their somatotopy is responsible for a sort of loose chemotopy; no other actual chemotopy is presently found in the centres. Further argumentation shows that spatial patterning is independent of any chemotopy and must not make us expect a chemotopy.

2.3.1 Tongue and Taste Nerves

In the early days of taste research, psychologists proposed a concept of differentially localized sensitivities of the tongue to the four basic qualities (Hänig 1901). Many authors have taken for granted the existence of tongue areas best sensitive to sweet, acid and so on. Collings (1974) tested this hypothesis.

She divided the human oral cavity into five different loci: front and side of the tongue, foliate papillae, vallate papillae areas, and palate. The local differences of sensitivities to the four standards obtained are very small, except when compared with the palate which exhibits much higher thresholds – the soft palate exhibits higher thresholds than the tongue, and the hard palate, a very poor sensitivity (Nilsson 1979).

The palatine taste buds are innervated by the greater superficial petrosal nerve (Miller 1977; Miller and Spangler 1982). The front and side of the tongue (fungiform papillae) are innervated by the C. T., the back of the tongue (vallate papillae) by the IXth nerve and the middle part (foliate papillae) by both nerves (Yamamoto and Kawamura 1975). Differences in perceived sensitivity can be related to innervation but within an area innervated by the same nerve(s) no topical organization is revealed. Gent (1979) and Bartoshuk et al. (1985) show "that 1/10 of the sensory field can produce taste intensities almost equal to those obtained with a whole mouth stimulation; good psychophysical functions ... were approximately similar for the front and rear ... ".

Oakley (1967, 1970, 1974) has shown that a mutual recognition mechanism between predetermined cells and fibres must occur during ontogenesis. This mechanism is probably related to axonal transport (Oakley et al. 1981, 1984; Sloan et al. 1983). The global sensitivity of a nerve should be correlated to the global sensitivity of the corresponding set of papillae. Localized sensitivities, if they can be observed reliably, should be at least correlated with different innervations. It is well known that the IXth nerve, which innervates the posterior part of the tongue, is more sensitive to quinine than to NaCl, whereas the C. T. which innervates the anterior part of the tongue is nearly equally sensitive to quinine and NaCl (Oakley 1970).

Interindividual differences may be responsible for the variance between author's assertions, from the beginning of the century, regarding which parts of the tongue are more sensitive to sweet or acid or bitter taste. As a matter of fact, each author reported on his own sensitivities or those of a very few subjects.

Collings (1974) found that in human subjects, contrary to what is generally asserted, quinine is best perceived by the tip of the tongue and the palate, not the vallate papillae. Collings submitted that perhaps the stimulation of the vallate papillae can reach the palate as well, so that high responses from vallate papillae could come from the stimulation of both regions.
Also, Cardello (1981) did not find circumvallate papillae to be more sensitive than fungiform papillae to quinine. In the rat in fact, neurophysiological data show that the IXth nerve in not better sensitive to quinine than the C.T. in an absolute sense, but with respect to sensitivities to NaCl.
These psychophysical results correlate well with Oakley's (1970) electrophysiological data showing that the glossopharyngeal nerve is not actually more sensitive to bitter tastants than is the C.T. The difference is relative and it is the ratio of amplitude of response to quinine to amplitude of response to NaCl which is higher for the glossopharyngeal nerve than for the C.T., because the former's response to NaCl is weaker than the latter's response to NaCl.
The higher sensitivity to quinine reported at the base of the tongue might well depend on great interindividual differences.

2.3.2 Chorda Tympani

Following Collings (1974) the response to sucrose could be different between the front and the side of the tongue. Should this be significant, it could then attest of an actual chemotopy within the C.T. innervation area. Yet data obtained at more elementary levels do not favour this hypothesis since single papillae, sensory cells or fibres are not specific, and their sensitivities are randomly distributed. The extension of C.T. fibre receptive fields can be presented as an argument against localized sensitivities. Miller (1971, 1974) has shown that C.T. fibres are branched and innervate several papillae. Oakley (1975) noticed that papillae partially innervated by the same fibre grossly present the same sensitivities; the distance between such interconnected papillae is 2 mm in the cat, but connected papillae can be found up to 18 mm apart. (Hence collaterals of one and the same fibre may extend over as much as 2/3 of the whole nerve territory.) If papillae which probably share the least dissimilar sensitivities can lie so wide apart, it is difficult to imagine that a topical organization of sensitivities on the tongue could exist.
Ishiko (1974) has electrophysiologically studied the sensitivity of the several branches emerging from the chordolingual junction between the C.T. (VII) and the lingual nerve (V), which innervate the anterior tongue. He successively cut each of the branches which innervate the different portions of the anterior tongue while recording the whole C.T. responses to stimuli applied to these various tongue areas. In the cat, it seems that the anterior branch, which is a low responsive branch compared with others, responds better to quinine, whereas the medial branch responds better to NaCl and the posterior branch better to HCl. It seems thus that "relative responsiveness of one lingual branch, differs with another, ... local taste specificity within the fields of cat's lingual nerve (chordo-lingual) depends not only on the size of taste receptive areas or the number of taste receptors but also on the differential sensitivity of a population of receptors" (Ishiko 1974, p. 353).

Ishiko shows that in the monkey the number of branches and collaterals increases so much that tongue areas overlap much more than they do in the cat. These areas are not only difficult to determine in one animal but apparently differ from one individual to another. This nearly prevents a topographical study. We can imagine that the same is true for the human as for the other primates; overlapping areas and interindividual differences may prevent any reliable sensitivity localization.

2.3.3 NTS and PTA Levels

Halpern and Nelson (1965) have shown the existence of three different tongue projection areas in the NTS: one where the anterior tongue projects exclusively, a more medial one where both anterior and posterior tongue project, and an even more medial third area where the posterior tongue projects exclusively. They stated that the responses to Q-HCl were larger, with respect to NaCl responses, in the composite area and even larger in the third one. This topography parallels the topical innervation of the tongue by the C.T. and IXth nerves. These authors concluded that a quinine solution, flowing from the tip to the base of the tongue, should produce a spatial temporal pattern in the NTS with an increasing magnitude with time. By contrast, an NaCl solution should produce successively decreasing magnitudes in the same areas, though with a similar temporal sequence. Halpern (1967) found a dorsoventral organization in the NTS as well; multi-unit responses to sucrose are found more dorsally, whereas QHCl and NaCl responses are found more ventrally. But single-unit recording leads to different results. Dickman et al. (1982) recorded NTS single neurons driven by both anterior and posterior tongue chemical stimulation. The neuron sensitivity often depended on the stimulated field. For example, "a neuron responded exclusively to HCl delivered on the anterior tongue and to quinine delivered to the posterior oral cavity". This is not exactly how we imagine, at first, that quality information should be extracted through topical organization. [3]

Convergence. Instead of a simplified pattern, we find more complexity due to convergence. Ogawa et al. (1982) and Ogawa and Hayama (1984) have shown that half of PTA neurons exhibit complex receptive fields including tongue and palate areas. Ogawa and Kaisaku (1980) reported that all combinations of converging inputs can be found in the rat NTS between lingual (V) and C.T. (VII), lingual and glossopharyngeal (IX), C.T. and glossopharyngeal and lingual nerves. Whether the chemotopy resulting from the different sensitivities of the VIIth and the IXth is lost or extracted by the convergence is a question worth asking.

[3] (Multi-unit recording shows the global response of an area the diameter of which depends on the electrode characteristics. Although it is well understood that the CNS integrates information originating from a number of units, one should not believe that multiunit recordings better represent the information that is taken into account by higher centres than single-unit recordings do. The group of units which are globally recorded by an electrode has probably nothing to do with the whole range of units which are altogether significant to the brain)

Although "PTA projecting neurons do not cluster in a particular region of the NTS but are intermingled with non-projecting ones", no quinine-sensitive neurons of the rostral NTS project to the PTA (Ogawa et al. 1980, p. 298). In the same rostral NTS area, Ogawa and Kaisaku (1980) could not find glossopharyngeal driven neurons projecting to the PTA. They suggested that glossopharyngeal projecting neurons should be found more caudally in the NTS, this area being more devoted to the IXth projections.
Although not easily understandable, there is a topical converging organization which does not obviously seem to rest on the grouping of qualitatively similar neural elements as shown by the study of single-unit receptor fields. The system seems to elaborate more and more intermingled patterns through a complex circuitry.

2.3.4 *Thalamus Level*

So far, researchers have not mapped actual chemotopy in the thalamus taste area. Ninomiya and Funakoshi (1982b) recorded responses to various chemical stimuli applied on restricted tongue areas comparatively. In 11 units out of 20, the receptive field did not vary when changing the stimulus quality. In five units the location and size of the receptive field changed with the stimulus applied, but the most sensitive region kept constant over such changes. In four units only, the most sensitive region of the receptive field depended on the stimulus applied.

2.3.5 *Cortical Level*

The projections of the Vth, VIIth bis and the IXth nerves are more or less superimposed on the cortical projection and largely depend on the animal (reviews by Norgren 1977; Yamamoto and Kawamura 1977; Yamamoto et al. 1980a). Moreover, Yamamoto et al. (1981b) showed that in the rat extralingual taste inputs (palate and epiglottis) share the same neocortical taste area with lingual inputs. Nonetheless, Yamamoto et al. (1980a, 1981c, d) found that the C.T. projection which lies anterior to the medial cerebral artery contains neurons sensitive to chemical stimulation and cooling, but not to warming or touch. They recorded such units while stimulating the tongue with the four standards and tentatively depicted a tendency to chemotopical organization (Yamamoto 1983; Yamamoto et al. 1984c). Even at the cortical level, where the information has its final shape, there is no obvious chemotopy – but a very loose, imprecise one.

2.3.6 *Quality Coding Without a Chemotopy?*

We already saw (Section 1) that neuron specificity does not obviously increase with the successive hierarchical levels of the sensory chain. Even at the cortical primary projection, neurons still present a spectrum of sensitivities. Moreover, similarly sensitive neurons are not obviously spatially grouped together. As a matter of fact, we can imagine that there is actually no need for a grouping of similar cells; a spatial pattern can be recognized although qualitatively similar cells are not spatially grouped.

Let us come back to the previously described analogy between a sensory pattern and a photograph. If the picture is black and white, it is the model of form recognition. Let us now compare our biological pattern to a colour photograph. Colour is then another dimension of the coding process, the dots of the picture appear nearly all of a different colour just as our cells nearly all exhibit a different sensitivity spectrum.[4] Dots of identical colour will not be grouped together to form the picture; instead, they will be properly dispersed all over the photograph, just as cells are, in order to form a pattern.

Looking for a grouping of identical cells at any level will give neither negative nor actually positive results. Green dots are grouped on a photograph to draw a tree, but you can see several trees on your photograph and all green dots will hence not be grouped together.

After being coded throughout a sensory chain, the information is supposedly decoded in view of a cognitive perception. The cortical primary projection area is thought to be the last step of the sensory chain where the information has taken its final shape ready for the final objective reading by more associative areas which will integrate it within more complex functions.

We can be sure now that, even within the cortical primary projection area, the quality is depicted by groups of cells and not single cells. What about secondary projections? Nobody has recorded responses to gustatory stimuli in such associative areas. But, considering the very large extent of cortex (nearly half an hemisphere) which must be removed in order to produce any deficit in a taste aversion paradigm (Benjamin 1963; Yamamoto et al. 1981b), we understand that this part of the brain is not stigmatic. (Pfaffmann has written that "after a cortical lesion, it is rather hard to show the rat has a deficit"; see Pfaffmann et al. (1977) for a discussion.) Instead it is holographic. Any piece of information lies everywhere in all functional units; that is to say the information is not lost when a part of the relevant brain area is physically cut off. The pattern itself, then, is reproduced many times, as in an hologram: breaking it into two parts gives twice the same original picture. Thus, in the holographic-associative parts of the brain, not only do the points of information lie in a pattern, but this pattern is reproduced many times in many groups of cells, the function of which is to put together information coming from various sensory systems, memory and so on.

On the other hand, I am not sure we can compare the organization of the primary taste projection to an holographic image to the same extent. But, it is not stigmatic either. A stigmatic projection would result in a point to point projection of the tongue onto restricted cortical areas, as the projection of a visual image is through an anastigmatic objective. This has not been demonstrated so far by researchers who have been actively looking for it. On the contrary, it seems that

[4] (Note, that there can be different types of responses: e.g. on or on-off responses, which are often found in the cortex though Yamamoto and Kawamura (1974) found them in the C.T., too; inhibitory or excitatory responses were found by Scott and Pritchard (1982) even in the C.T. nerve. These electrophysiological characteristics do not appear to be associated with any particular quality)

any information lies in a group of cells which are scattered over the cortical area which is already somewhat comparable to an holographic system.

Hence, nowhere is there to be found a reduction of information through a funnel, so that one single cell should hold a meaningful message. Instead, groups of cells are the functional units. How, then, to apprehend them experimentally?

This pattern coding, which scientists have guessed the existence of through unitary recordings, can obviously neither be approached nor visualized using conventional electrophysiological techniques, since we are not able to record a lot of units at a time.

Looking towards the future, we can ask ourselves, from where will new heuristic concepts emerge? With what kind of experimental data will we tread a step forward in understanding brain coding mechanisms? We should tentatively suggest that the answer lies in experimental access to the whole-system synthetic response and not in the analytic processing of the responses of single units. In other words, now that we have understood pattern coding, how can we use our knowledge and what kind of experimentation can we perform to go a step further from this basis? In a lecture given in Philadelphia in 1983 (reported in Erickson 1984), Erickson presented most of these same ideas. However, after offering arguments and explanations to describe how neurons form patterns, he quoted Mountcastle and Sperry worrying about the method for obtaining huge populations of fibres and their responses in order that the experimenter may be able to deal with patterns. To which worries, Erickson answers: "this is not a problem at all. Populations of data are presented in the literature ... they are just not presented in the right form to show the population responses." I agree with his second sentence, but not with the first one. If we only want to discuss theoretically and exhibit patterns, we have as much data in the literature as we need; it is just a matter of occasionally reinterpreting them when necessary, when possible. But, if we want to use these responses to understand a little bit more about chemoreception, we have to deal with many responses of many neurons to many stimuli – which is a great deal. Figure 1 shows responses of intrapapillar single units recorded by us in the hamster. These two matrices are subsets of results (Faurion et al. 1982, 1983). Factor analysis of responses to only six stimuli allows some units to be grouped together; their profiles of responses are identical. The same treatment of the same unit responses to 10 stimuli discriminated all units scattered over a continuum. It is necessary to use enough stimuli to characterize unit sensitivity profiles and, as a corollary, enough units to characterize the sensitivities represented by a stimulus.

Specialists of data treatment argue for a factor of ten between the smaller dimension of the matrix (for us, units or stimuli) and the number of factors responsible for the matrix data variance (for us, independent mechanisms for coding our data, for example, the number of independent channels which combinatorily code qualities). If we have just a small number of peripheral receptor mechanisms, for example, smaller than 10 (and this is not sure), it means that we will have to record a matrix of at least 100 units by 100 stimuli to extract them as factors, using factor analysis. What is the representativity of 10 stimuli with respect to this problem? Very poor. Several dozens of sweeteners are readily available now; several dozens of bitter compounds, too, are available. Even if I can envisage working with dozens of compounds on more than a hundred subjects (over years), I am not sure

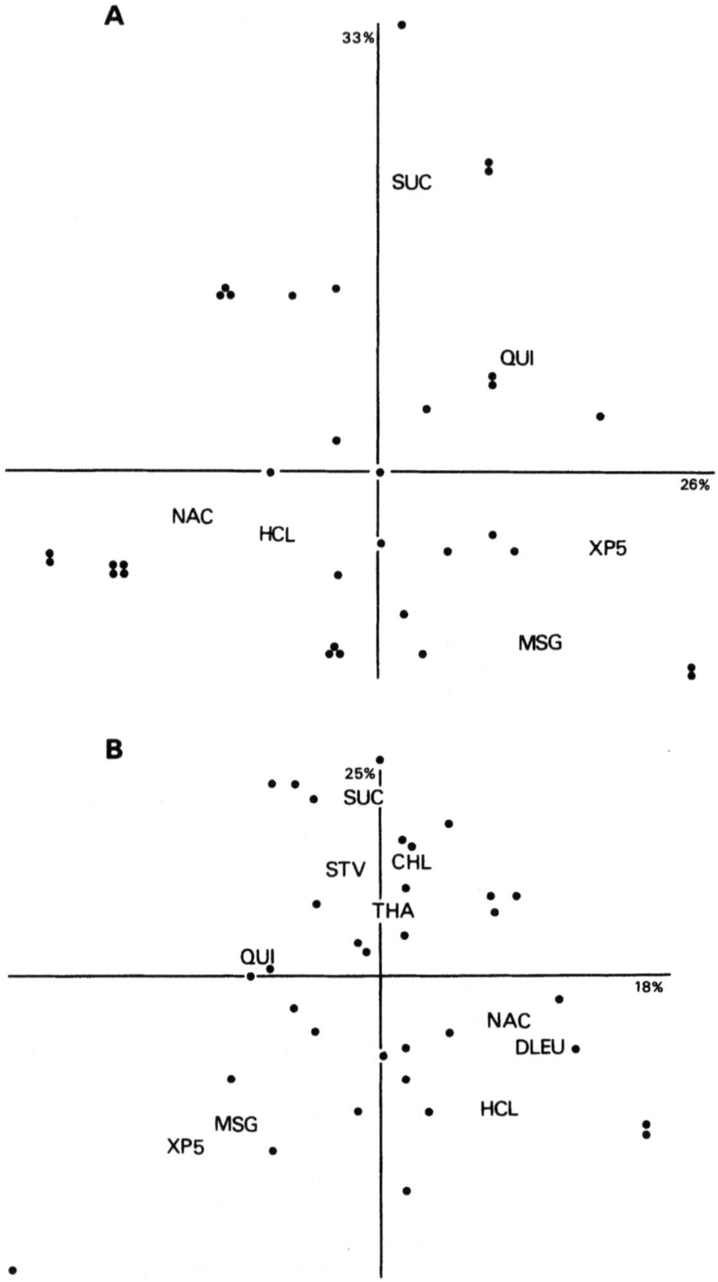

Fig. 1 A, B. Factor analysis of single units recorded in the pore of the taste bud using a glass micropipette, for six and ten stimuli. Units which are grouped in **A** responded similarly to all six stimuli (responses were coded by 0 and 1: 0 if the unit had never responded to the stimulus, 1 if it had responded once or several times with a criterion of 2 s.d. above the mean activity during the 1 min preceding the stimulus). In **B** these units are no longer grouped, since their response profiles were distinguished by the four other stimuli. (Faurion and Audouin, unpublished data)

we can expect to record enough unit responses to many stimuli using conventional electrophysiology.

2.4 Complex Multimodal Sensory Image Rather than Pure Taste Image?

In everyday life, and even in the laboratory, it is easily noticed that the sensory image is not perceived as a pure taste image; if you taste a sample solution, you can appreciate its temperature, texture (astringency, eventually), and taste, and you may even not be conscious that what you call taste are in fact a flavour including gustatory and odorous components.

2.4.1 Psychophysical Observations Correlated to the Hypothesis of a Multimodal Sensory Image

The effect of temperature on taste responses in man has been studied by Pangborn et al. (1970), McBurney et al. (1973), Moskowitz (1973), Paulus and Reisch (1980), and Bartoshuk et al. (1982) and proven to be clearly significant. For example, the intensity perceived when tasting a sucrose solution is significantly higher if the solution is warmer.

The independence and integration of olfaction and taste have been discussed by Mozell et al. (1969), Murphy and Cain (1980), Hornung and Enns (1984). Indeed, it is a very impressive experience to dissociate taste from retronasal olfaction by means of an airflow entering through the nostrils. Intensity data are different, particularly near threshold (Faurion, unpublished data).

Meiselman et al. (1972) studied the effect of flow rate on taste intensity response in man. In accordance with his results, I found that below a rate of 40 ml/min, in a continuous flow procedure, a subject could not distinguish between water, 1 *M* sucrose or 0.5 *M* NaCl, whereas the quality recognition was complete over 60 ml/min (unpublished data). The mechanical effect due to high flow rates is an important feature of the "taste" sensation.

Lawless and Stevens (1984) compared taste intensity evaluations with and without rinsing the subject's tongue with capsicum, an irritating substance which is supposed to produce a trigeminal sensation.

Cardello (1981) studied the quality perceived by subjects during tactile and electrical stimulation of single papillae, and proposed to explain the consistency of quality responses provoked by both tactile and electrical stimulation by cross-modal associations. (Although consistent, judgments may differ qualitatively from responses to chemical stimulation of the same single papilla.) He quotes Marks (1978), who already presented evidence of cross-modal associations in sensory systems, including taste.

The commentaries of Cardello's subjects are striking. One "subject said that she could relate the *sensations* of the chemicals but the *taste* was not the same". Another said "the tingling of this salty taste (it was an electrical stimulation) is like the tingle of chemical salts at high concentrations". This suggests that non-gustatory sensations are interpreted as gustatory qualities. Let us note that this result comes from single-papillae stimulation, that is to say that only a small part

of the sensory taste system is activated, and the task is difficult for the subject. In these conditions, the elicited neural pattern is so poor that high concentrations of chemicals are needed and many errors are made by the subjects.
Hence, it may be suggested that if there are obvious components of other modalities than taste in the elicited neural pattern at the level where quality is not surely discriminated, then it is reasonable to think that the same components exist when a better sensory image is obtained with whole tongue stimulation. The taste feature is then well recognizable. If the subject's task is to recognize taste quality, he will no more be tempted to speak of "tingling" and so on because he will neglect this information, and he will even not be conscious of it. The pertinent information wipes out other components which may be the less faint clues when the testing situation is made difficult.

2.4.2 Electrophysiological Data

Convergence between the VIIth, Vth, and IXth nerves on NTS neurons has already been reported. Must we consider the existence of a pure taste cognitive image or, on the contrary, of a multimodal one containing thermal, mechanical and chemical (taste + olfactory) information? Stimuli which are not specific to a given modality are usually considered as inadequate. Touch and temperature could be considered as inadequate stimuli for the taste system. Indeed Pfaffmann (1941, p. 252) considered touch as an inadequate stimulus for the chorda tympani.
The CNS obviously integrates primary sensory inputs in non-specific projection areas, but the question arises here whether these thermal, mechanical and chemical tongue inputs are actually isolated somewhere. In other words, which is the hierarchical level where one and the same pattern signals that a stimulus is salty, warm, astringent and smells of garlic all at the same time? Is this multimodal pattern the result of integrative functions in cortical secondary projections, or is it already constituted at lower levels of the sensory chain? In this latter case, is it obtained through convergence of separated sensory inputs or by multimodal sensitivity of afferent nerves? Conversely, if a part of the peripheral afferents transmits some polymodal sensory information, are these modalities properly separated in the primary cortical projection? Reviewing the literature, we shall see that the answers are not quite so clear and that they depend on which modality is being considered.

2.4.2.1 Olfactogustatory Convergence and Tongue Stimulation with Odorous Components

The case is simple for the olfactory part of the chemosensory image: since peripheral organs are separated, we are dealing with a convergence which has been demonstrated to exist at least in the NTS. Van Buskirk and Erickson (1977) found 20 neurons out of 30 in the gustatory NTS that responded to odorants as well as gustatory stimuli. Giachetti and McLeod (1977) found neurons in the thalamus of the rat which responded to electrical stimulation of the tongue and lateral olfactory tract. This convergence may concern the trigeminal innervation of the

tongue and not the gustatory modality (chorda tympani). Hence, it should better be called an olfactolingual than an olfactogustatory convergence.
More surprisingly, Kashiwagura et al. (1977) demonstrated that the taste nerve of the frog (glossopharyngeal) is sensitive to odorants as well as to gustatory stimuli. Could it be that taste and smell are not well separated during food mastication? Could a part of the odorous components of food stimulate tongue receptors as well as olfactory receptors at the same time? If so, neither a pure olfactory nor a pure gustatory image would ever be associated with food intake.
This is the case, for sure, for the fish which can "taste" and "smell" amino acids or other chemicals diluted in its water bath (Caprio 1978; Marui et al. 1983a, b). Moreover, one can ask whether receptor sites could be similar for taste and olfaction or whether at least some of our olfactory and gustatory receptor sites are the same.

2.4.2.2 Thermal and Mechanical Sensitivity of Taste Neural Elements

The anterior part of the tongue is innervated by two so-called specific nerves, the chorda tympani (VII bis) for taste, and the lingual (V) for somesthesis. Whereas the posterior tongue is innervated by only one mixed nerve, the glossopharyngeal (IX).

Glossopharyngeal Nerve. In the sheep, fibres of the IXth nerve have been shown to respond far more to chemical stimulation if the receptive field is mechanically stimulated (Iggo and Leek 1967). At least for the glossopharyngeal part of taste, touch and taste are encoded together in the same pattern.

Chorda Tympani. The thermal sensitivity of the chorda tympani is well documented by Sato M. (1963), Ogawa et al. (1968), Yamashita et al. (1970); Erickson and Poulos (1973) have shown that single fibres are nearly all sensitive to cooling or warming or both. Moreover there is a correlation between sweet sensitivity and warm sensitivity.
As far as touch is concerned, Oakley (1975) said that "pressure on a fungiform papilla often excited a taste (C.T.) fiber. The mechanical sensitivity aided in the task of localizing fungiform papillae responsive to taste."
Biedenbach and Chan (1971) estimated conduction velocities and latencies of afferent fibres responding to mechanical stimulation of the tongue in the lingual nerve and in the C.T.
Authors working at determining chemical receptive fields of units recorded in a central structure commonly begin by mechanical stimulation for a first localization. It is presently very difficult to conclusively state that mechanichal stimulation is an inadequate stimulus for C.T. fibres. A very gentle approach with a glass micropipette to any papilla will elicit spikes in nearly any topically related fibre.

Central Projections. In the medulla, the converging projections of the C.T. and lingual nerves have been studied by Beckstead and Norgren (1979), Whitehead and Frank (1983) and Hamilton and Norgren (1984). The latter authors thoroughly investigated the NTS projection of the Vth, VIIth, IXth and Xth nerves. Ogawa et al. (1984) showed a convergence of tongue and palate chemosensory and mechanical inputs.

In the brain stem, many taste neurons are sensitive to mechanical stimulation as well as taste: 13 out of 21 for Makous et al. (1963), 14 out of 107 for Ishiko and Akagi (1972), 84 out of 142 for Owaga et al. (1981) in the NTS. Data concerning PTA are contradictory: 24 out of 32 are sensitive to touch as well as taste for Ogawa et al. (1981), while none are for Scott and Perrotto (1980).

Travers and Smith (1984) show a co-variance between sucrose and warming and another covariance between cooling and NaCl stimuli.

In the thalamus, Benjamin (1963), noticed that 55% of neurons in the cat and 85% in the monkey responded only to gustatory stimuli, the rest responded also to thermal stimulation; he did not find taste units responding to mechanical stimulation.

In the gustatory cortex, only a small part of the area which is anterior to the medial cerebral artery contains neurons responding to taste and not to touch (Yamamoto et al. 1980a, 1981c, d); however, the pure taste cells which have been recorded there are still sensitive to cooling of the tongue. Conversely, the C.T. projecting neurons localized in the area posterior to the medial cerebral artery are sensitive to touch and temperature (cold and warm) as well as taste (Landgren 1957; Yamamoto et al. 1980a, 1981d).

As far as sensitivity of the anterior two thirds of the tongue is concerned, we can thus suggest that taste and thermal modalities, at least, are probably encoded together in the same pattern throughout the taste system. The taste sensory image is at least a gustatory and thermal sensory image. There is a second complementary sensory image, however, which covers taste, thermal and mechanical modalities in the cortical primary projection, and in the part of the gustatory cortex which is posterior to the medial cerebral artery, the sensory image covers taste, thermal and mechanical modalities.

Whether olfaction is partially separated or completely confused within these primary projections has never been tested. Nobody has recorded such gustatory responding cells in response to olfactory stimulation, but the existence of such olfactogustatory converging cells is probable. Giachetti and McLeod (1975) recorded single units responding to odours in the area which is posterior to the middle cerebral artery, which, at least, responded to electrical stimulation of the tongue (somesthesis). Shipley and Geinisman (1984) recently confirmed the validity of this hypothesis by showing an anatomical convergence of gustatory and olfactory pathways onto the insular cortex in the mouse.

The existence of a multimodal sensory pattern seems well documented at any level on the sensory chain. This fact has a methodological consequence. When an experimenter drops a chemical on to the tongue of an animal, several sensory modalities are all concerned in the same electrophysiological response. Comparing the responses to chemicals A and B requires that neither temperature nor mechanical stimulation be added to gustatory stimulation. As a matter of fact, it is of nearly no use to subtract the response to a control mechanical stimulation, such as stimulation with a rinsing liquid, since we do not know whether chemical, mechanical and temperature responses are additive. Hence, it is better to suppress spurious stimulation, and use a continuous flow system, for example.

2.5 Cognitive vs Non-Cognitive Pathways?

Some of our neurophysiological knowledge necessarily comes from animal experimentation. Electrophysiological recordings must be obtained from animals and, behavioural experiments are thus necessary correlates of electrophysiological recordings. However, we cannot directly compare physiological data obtained on the animal with our own human perceptions. Since such behavioural experiments necessarily deal with behaviour-related mechanisms as well as sensory mechanisms, we have to distinguish which part of the brain is related to behaviour and which is actually related to sensory mechanisms.

2.5.1 Brain Stem

The brain stem is the lower common part of the brain which gives way to two separate pathways in which gustatory responses can be electrophysiologically recorded. The sensory information is sufficiently clear in the brain stem to evoke behaviour reflexes such as a quinine rejection and sucrose acceptance. Grill and Norgren (1978) have shown that chronically decerebrated rats are just as capable of mimetic reactions related to quality as normal rats are. Such mimetic reactions have been observed even in the human anencephalic neonate by Steiner (1973). Such quality-dependent reactions prove that quality is depicted clearly enough in the brain stem to evoke feeding behaviour reflexes, although there is presumably no cognitive image. In this respect, we can understand that bitter taste rejection and sweet taste acceptance seem to be "a given", and not culturally acquired. [It is however possible for an animal to acquire bitter preferences. The chimpanzee "Sarah Premack" likes black unsweetened coffee and beer and prefers a moderately bitter solution of 10^{-4} *M* Q-HCl mixed to 3% sucrose to a 3% sucrose solution without quinine; see Dr. Rozin's intervention in the discussion of Glaser's paper, in Glaser (1979).]

Up to this level, there is place for all medullary and brain stem reflexes, integrating chemoreception from stomach and jejunal origin through the vagal nerve, and including feeding related reflexes (Norgren 1981; Giza and Scott 1983; Hermann et al. 1983). But, this reflex activity must not be confused with preference behaviour. The rat accepts or rejects a solution according to a very simple schedule which is not modifiable by any positive or negative reinforcement. On the other hand, descending modulations, from lateral hypothalamus, may affect afferent activity in the NTS of rats (Matsuo and Kusano 1984; Matsuo et al. 1984). Moreover, conditioned taste aversion, involving hindbrain structures, has been shown to modify the afferent activity of the NTS (Chang and Scott 1984).

In the PTA of the rabbit, Schwartzbaum and Block (1981) have found units responding to hedonic properties of tastants, units corresponding to movements and, units responding to both movements and hedonic properties of tastants.

2.5.2 Cognitive Pathway

It is usually considered that the cortical projection via the thalamus is responsible for cognitive functions. Cortical lesions specifically disrupt taste aversion learning, but not preference behaviour. This has much more to do, however, with the mechanisms of acquisition and maintainance of taste aversion than with sensory information.

When the gustatory neocortex is removed, taste agnosia results but, "the loss appears to be specific to the memorial significance of taste stimuli and thus can be labeled an agnosic disorder" (Kiefer et al. 1984).

Although the cerebral cortex is the best support for cognitive functions, no experimental evidence has ever been reported on pure cognitive sensory images.

When rats are decorticalized, they reject solutions of any taste quality. At first sight, this appears to be in contradiction with Section 2.5.1., where it is shown that if the thalamus has been destroyed, too, the brain stem is sufficient to elicit the necessary reflexes, sucrose acceptance and quinine rejection. But the thalamus is controlled by a corticofugal feedback loop: 20% of thalamic cells are inhibited during 60 ms by electrical stimulation of the cortex, and 40% are inhibited for 10 ms and facilitated for about 60 ms (Yamamoto et al. 1980c). We then understand that in decorticalized animals still possessing a thalamus, the expression of the brain stem reflexes is not free but under thalamic control, whereas the cognitive image is not elaborated enough for this control to be properly oriented.

2.5.3 Non-Cognitive Pathway

The second branch of the Y emerging from the PTA projects to the hypothalamus, amygdala and bed nucleus of the stria terminalis (Schwartzbaum and Block 1981; Block and Schwartzbaum 1983).

The lateral hypothalamus is a locus of convergence for olfactory, gustatory, tactile and visceral inputs (Fisher and Almli 1984).

In the rat hypothalamus, Norgren (1970) has found units which respond to sucrose and induce self-stimulation behaviour, and units which respond to quinine and conversely do not induce self-stimulation behaviour.

Lesions of the amygdala interfere with feeding behaviour (Fonberg 1969; Nachman and Ashe 1974) and specifically disrupt taste preference among other affective responses (Kemble and Schwartzbaum 1969). Azuma et al. (1984) recently showed that taste information is processed in a hedonic dimension in the amygdaloid neurons.

Moreover, an amygdalocortical projection to the gustatory neocortex seems responsible for the taste aversion retention (Lasiter 1982; Yamamoto et al. 1981a, 1984a), but not for taste aversion acquisition. "These interconnections may play some role in association of taste-related cognitive processes with feeding behaviour" (Yamamoto et al. 1984a, p. 23).

It seems that the structures involved in preference and reward should belong to this second pathway, including hypothalamus, amygdala and substantia inominata, where Rolls et al. (1976), and Rolls and Rolls (1977) have found units responsive to the sight of nutrients when the animal *is not* satiated.

Schwartzbaum and Block (1981) present the hypothesis that the PTA works as an interface for structures like amygdala and substantia inominata to modulate motor mechanisms involved in ingestive behaviour. The PTA's hedonically responding neurons may lead to hedonic evaluation of sensory stimuli. The ventral forebrain structures may integrate this kind of information with other kinds, for example, gustatory or visceral information, eventually leading to modified behavioural reactions. The relationships between the hedonic evaluation, the behavioural reaction and the cognition are far from being well understood and the distinction between cognitive and non-cognitive functions is controversial. Is hedonic evaluation a process belonging to cognitive functions? Probably not necessarily since Grill and Norgren's rats were decerebrated when rejecting or accepting a stimulus, but this is still a matter of opinion.

As far as hedonism is concerned, one can wonder whether it is only strictly related to diencephalic structures such as the hypothalamus, or is it related to cognitive brain areas as well? Indeed Burěsová (1978) and Yamamoto et al. (1981a, 1984a) demonstrated functional connections between the neocortex and the ventral forebrain (amygdala, hypothalamus). Do we like a taste because we recognize its quality, or do we independently like it and recognize its quality? Could we like a taste sensation before we are aware of its quality? The degree of interrelationship between cognition and preference is presently not really known.

Some degree of independence between cognition and hedonism is likely. However, we now have one argument that shows that the pattern of C.T. fibre responses may change after the stimulus has been made aversive to the animal (Scott and Chang 1984).

Behavioural tests performed to study the sensitivity to gustatory stimuli do not readily show which part of the brain is actually involved in the observed behaviour. Different experimental procedures may involve completely different taste projections. As a result, animal behaviour may be a difficult tool to deal with when studying peripheral chemoreception, although it is sometimes a necessary correlate of electrophysiological data.

3 Part B: Present Understanding of Physiology of the Sweet Taste

In this part of the paper, we shall focus on sweet taste physiology. Sweet taste is not a peculiar taste quality as far as neural coding is concerned (the relevant aspects of which have been treated in Part A). Sweet taste shares with bitter taste the peculiarity of being elicited by a wide number of organic molecules, whereas salt or acid tastes are not. Also sweet taste differs in the way that it is related to medical disorders; this is probably the origin of the strong research effort to understand chemoreception mechanisms related to sweet taste. The last two assumptions help to explain why a model of a receptor site (Shallenberger and Acree 1967, 1971; Kier 1972; Hough and Edwardson 1978) has been proposed for sweet taste only. A first attempt (Kubota and Kubo 1969) followed by Wieser and Belitz (1975) and Belitz et al. (1979) among others, to characterise a receptor site to bitter taste did not give rise to a clearcut model. Sweet taste is the best known and a more interesting model for "fundamentalists" than other tastes, as far as the acceptor mechanism is concerned.

We shall develop converging arguments here which suggest that sweet taste is not a unique quality. These arguments are found in the multivariate aspect of either species (Section 3.1), individual (Section 3.2), or even single neuron responsiveness to a set of sweeteners (Section 3.3). Profiles of quantitative responses invariably show that sensitivity to a sweetener depends on the subject (or perhaps, the neuron) as well as on the sweetener considered. The variation among subjects (or neurons) of sensitivity to various sweeteners is found to be multidimensional. To account for such multiple sweetness sensitivities, we must consider the existence of several membranous acceptors. Arguments in favour of this hypothesis emerge from cross-adaptation results (Section 3.4) and experiments with inhibitors (Section 3.5).

The theory of the existence of several receptor sites can be related to recent biochemical findings (Section 3.7).

In some respects, sweet taste can be compared to bitter taste which depends on two receptors, at least, one of them being more or less specific for PTC and related compounds (Fox 1932; Hopkins 1942; Kalmus 1971; Lawless 1980). Lush (1981, 1982) and after him Harder et al. (1984), found among mouse strains genetically related differences of sensitivity to sucrose octaacetate and strychnine hence two receptor sites at least to bitter tastants.

A second concept will be discussed here which appears in close correlation. Most of the same arguments converging to suggest the non-uniformity of sweetness sensitivity also suggest that the whole range of tastes is not convincingly described by four qualities; on the contrary, a wider set of sensitivities is to be considered.

3.1 Interspecies Differences

3.1.1 Non-Uniformity of Sweet Taste

Stimuli which are sweet to man exhibit different stimulatory properties in different animals, either on a qualitative or a quantitative basis. Beidler et al. (1955) showed that the hamster and the guinea pig are far more sensitive to sucrose than is the rat. Hellekant et al. (1981) recorded C.T. responses to six sweeteners in five primate species. As can be seen in Table 8 none of the primates tested was sensitive to aspartame, whereas *Macaca fuscata* is (Sato et al. 1977c). Hellekant (1975) did not find C.T. responses to the sweet protein thaumatin in either dogs, guinea pigs, hamsters, pigs, rabbits or rats, but only in some higher primates, such as *Cercopithecus aethiops* (Hellekant et al. 1976).

On the basis of behavioural experiments, Glaser (1979) and Glaser et al. (1978) suggest that thaumatin sensitivity divides the primate order into two groups, sensitive and non-sensitive, but, among thaumatin non-sensitive primates, some can perceive monellin, a sweet protein very similar in taste quality and efficiency to thaumatin in man. Among primates, monellin can elicit either rejection, accep-

Table 8. Whole chorda tympani responsiveness of three primates to three sweeteners

Primate	Glycine	Acesulfame-K	Aspartame	Reference
Tupaia glis	+ +	+	0	Hellekant et al. (1981)
Callithrix j. jacchus	+	+ +	0	Hellekant et al. (1981)
Macaca fuscata			+	Sato et al. (1977)

Table 9. Preference behaviour of some primates to different sweeteners. (After Glaser 1979)

Primate	Sucrose	Monellin	Thaumatin
Cercopithecus aethiops	+	+	+
Lemur mongoz	+	+	0
L. variegatus	+	$\frac{1}{2}$+	0
L. catta	+	0	0
Saguinus midas niger	+	0	0
S. oe. oedipus	+	0	0
S. fuscicollis nigrifons	+	$\frac{1}{2}$+	0
Callithrix j. jacchus	+	0	0
Cebuella pygmea	+	$\frac{1}{2}$+	0

+, prefers sweetener to water; 0, does not prefer sweetener to water; $\frac{1}{2}$+, intermediary reaction

tance, intermediary reaction or even no reaction at all (Table 9). Moreover, Glaser (1979) shows great behavioural differences across species towards a variety of monosaccharides and other sweeteners. As a matter of fact, preference cannot be considered as a measure of sweetness sensitivity, but, by reference to sucrose which is always preferred to water by all primates, we can understand that rejection behaviour or neutral reaction elicited by different sweeteners proves that these stimuli act differently on the considered organisms, whereas no difference between behavioural responses does not prove that the two stimuli elicit identical qualities for the organisms. Thus, although preference and quality perception are independent, we may conclude that there is a qualitative difference when there is a behavioural change from preference to rejection or even to indifference. Let us note here some unexplained discrepancies between electrophysiology and behavioural data. Hellekant (1975) has not found electrophysiological responses in the rat or hamster whole C.T., although the rat is able to distinguish a thaumatin solution from water and prefers it (Brouwer et al. 1973); moreover, thaumatin seems to reinforce the sweet taste of sucrose in the rat (Du Villard et al. 1980). Thaumatin-sensitive receptors could be innervated by other nerves than C.T., or we must discover a bias either in the electrophysiological or behavioural experiments.

The sweetness efficiency of some carbohydrates ranks in the following order for *Homo sapiens*: sucrose > fructose > lactose > glucose > arabinose > galactose > raffinose. Raffinose is barely sweet to man, but is the most efficient sugar for *Cebuella pygmea* and *Saguinus* (primates), whereas galactose, which is sweet to man, is preferred to water by *Cebuella* and rejected by *Saguinus* (Table 10). Such discrepancies are found for other sweeteners. Xylitol is sweet to man, preferred to water by some primates and rejected by others. Saccharin is preferred by some primates, rejected by others, and yet others seem not to perceive it. The differences of preference for various sweeteners do not co-vary; they do not vary in co-dependence across species.

Should the observed differences be summed up by clustering the species in several coherent groups, we could conclude that the receptor sites accepting molecules

Table 10. Preference behaviour of some primates to different sweeteners. (After Glaser 1979)

	Sucrose	Raffinose	Arabinose	Galactose	Xylitol	Saccharin
Homo sapiens	+	ε	+	+	+	+
Cebuella pygmea	+	+	$\frac{1}{2}$+	+		−
Saguinus midas niger	+	+	−		+	−
Tupaia glis	+				−	+
Callithrix j. jacchus	+				+	0
Galago senegalensis	+				$\frac{1}{2}$+	−

+, prefers sweetener; −, rejects sweetener; 0, neutral reaction; ε, just noticeable response (symbols are descriptive for sensitivity in the human)

which taste sweet to man are different across these various groups and unique inside each group. On the contrary, such grouping is impossible since sensitivity differences do not co-vary across the various species studied (Table 10).

Haefeli and Glaser (1984) measured xylitol thresholds in nine primates, among them man. The correlation coefficient we can calculate on the nine pairs of figures presented in their data is −0.1. This indicates non-covariance between sensitivity to xylitol and sucrose across the nine primate species.

As every species exhibits a different spectrum of sensitivities, either (a) there are as many receptor sites as different species, or, (b) each species possesses several receptor sites to encompass a range of sensitivities. This multiple equipment should then be slightly different from one species to another on a qualitative and on a quantitative basis. These receptor sites may be partially common to all species and quantitatively irregularly distributed among all species.

In the first hypothesis, the primate *Saguinus midas* niger (neither sensitive to monellin nor to thaumatin), the primate *Lemur mongoz* (sensitive to monellin, but not thaumatin), and the human, sensitive to both proteins should each have a different receptor site.

The corollary of this hypothesis is a high selectivity[5] of the receptor site; there is the rub. It is difficult to imagine a highly selective receptor site giving rise

[5] We shall use selectivity in its first sense which is independent of the sense of specificity. The receptor selectivity is the receptor's ability to select some ligands among others to which it is not sensitive. It is useful to distinguish these two concepts in a range of medium specificity. A receptor which is not strictly specific since it can bind three ligands is nonetheless selective since it cannot bind a fourth one. Two receptors can select different sets of ligands and thus exhibit different selectivities. Of course, if the binding is not at all specific there is no selectivity either. In this acceptance selectivity is more qualitative than specificity which is more quantitative; it is only phenomenological, being related to stimuli which appear efficient without any reference to adsorption mechanism

to a wide spectrum of sensitivities, i.e. a selective receptor cannot be sensitive to so many different sweeteners. As a matter of fact, thaumatin, aspartame, saccharin and sucrose are all stereochemically different (whereas sucrose and raffinose may be considered as less different) and a chemoreceptor accomodating several of them cannot be selective. A receptor selecting many ligands should be very poorly specific, i.e., would bind nearly any chemical structure, and hence, would have very small chances to actually select some ligands among others. Therefore, the second hypothesis is more satisfactory. The existence of several receptor sites sensitive to sweet stimuli allows any combinations of sensitivities, e.g. a group of stimuli can interact with receptor sites 1, 2, 3, or receptor sites, 1 and 2, or 1 and 3 etc. Moreover, if we consider that receptors should not be specific, but only present a higher affinity for some given sweeteners than for others, relative differences in response magnitude could be accounted for. For example, Hellekant et al. (1981) show that *Tupaia glis* and *Saguinus midas* niger give a higher magnitude C. T. response to glycine than to acesulfame-K, whereas *Callithrix j. jacchus* and *Galago senegalensis* give a higher response to acesulfame-K than to glycine. We can suggest that there are different relative amounts of acesulfame-K or glycine preferentially sensitive receptor sites in the considered species. Studying the relative magnitude of responses among species is now popular. Iwasaki and Sato (1984b) compared the relative sensitivity of macaque monkey, mouse and rat to saccharin and sucrose. Data from different reports are normalized to NaCl response and seem non-covariant. Ninomiya et al. (1984) compared gustatory neural responses and found differences among three strains of mice. The magnitude of the response depended on the stimulus and the strain.

The hypothesis that several non-specific acceptors exist in various amounts among species is a rather satisfactory way of explaining different sensitivities of species which are closely related to one another.

3.1.2 How Many Qualities?

Preference or rejection observed in behavioural tests does not indicate whether the animal perceives the various sugars of synthetic sweeteners of the same quality as sucrose. Data obtained from a taste aversion paradigm give an idea of what the stimulus tastes like to the animal. We do not know whether sucrose is "sweet" and NaCl "salty" to hamsters, but we know that Na cyclamate tastes like sucrose and like NaCl to hamsters (Nowlis et al. 1980), but like quinine-HCl and NaCl to gerbils (Jakinovich 1981). Once more, interspecies differences are apparent.

More surprisingly, Jakinovich (1981) has shown that chloroform, neohesperidin dihydrochalcone (sweet to man) and naringin dihydrochalcone (bitter to man) give electrophysiological responses in the gerbil although they taste like neither NaCL, HCl, sucrose nor quinine-HCl in behavioural tests for the same animal. Similarly, Nowlis et al. (1980) have found that aspartame presents none of the four basic tastes to the hamster in a conditioned aversion experiment, whereas we have recorded some single units responding to aspartame in the hamster papillae (unpublished data). Hence, stimuli which elicit an electrophysiological response but no common perception to any of the four standards in the same animal suggest that four reference qualities do not describe the whole taste modality.

In man, psychophysical data from Schiffman and Erickson (1971) have already shown that various chemicals could not fall within any one of the four qualities, but in between salty, sour and bitter. Yoshida and Saito (1969) and Yamaguchi (1967, 1979) using similar methods have shown that "Umami" (the taste of monosodium glutamate) is qualitatively different from any of the four basic tastes.

It seems that we have to deal with multiple sensitivities within sweet taste and more than four qualities for the whole range of taste. Hence, taste sensitivity and particularly sweet taste appear more complex than conventionally described. What should be the underlying correlates of such a complex space[6] at the receptor site level?

Should we assume, first, that there is one receptor site for each of the four standards (although it is not clear that acid taste being related to pH depends on a chemoreceptor protein)? Could these four receptors account for multiple sensitivities within sweet taste? Yes, but, they could not account for more than four qualities within the whole range of taste. Should we assume, on the other hand, that more than four receptors do exist (at least two receptors are already considered for bitter taste)? Could each of them participate in more than one quality sensation? The question remains open. We shall develop further relevant data and arguments.

3.2 Interindividual Differences in Sweetness Sensitivity

Non-covarying interindividual differences have been found as well as interspecies differences in sweetness sensitivity.

3.2.1 Interindividual Differences in the Monkey

One can see discrepant sensitivity differences among monkeys of the same species in the data of Sato M. et al. (1977c). One of the tested monkeys was more sensitive to aspartame than to stevioside. Conversely, another one was more sensitive to stevioside than to aspartame; whereas others responded equally to both sweeteners. Yet, they all belonged to the same species *macaca fuscata.* It must be emphasized that individual electrophysiological and biochemical data varied accordingly. The protein material extracted from the tongue of the animal which was more sensitive to aspartame, more strongly bound to aspartame than to stevioside in vitro.

---

[6] "Taste space" is short for spatial representation of taste, i.e. the figurative way we spatially represent distances between stimuli according to their similarity. This similarity may be measured in terms of perceived quality or neuron sensitivity. Each distance between each pair of stimuli is weighted as a function of the intersubject's measured variance

3.2.2 Interindividual Differences in the Human

We ourselves have reported experiments using quantitative psychophysics. We measured parameters of the sensitivity of each subject for each stimulus, thus constituting a rectangular data matrix (subjects × stimuli) either at threshold or at suprathreshold levels.

First, we measured sweetness recognition thresholds of 91 subjects to seven sweeteners and found no predictability rule within the data set. Sensitivity to paired sweeteners across subjects or sensitivities of paired subjects across sweeteners did not co-vary. All profiles were different, whether they were considered across subjects or across stimuli.

Second, we measured the variation with concentration in perceived intensity of sweet taste for 12 stimuli and 9 subjects. The slope obtained through linear regression of perceived intensity vs concentration does not co-vary either among sweeteners or among subjects, except between some of the 12 stimuli which thus appear not very different from one another, for example, sucrose and saccharin (Faurion 1982; Faurion et al. 1977, 1980).

In these analyses, we did not pool the subjects' data . On the contrary, we analysed and extracted significant information from the variance between individual data. Except for the threshold experiment, each subject's response to each stimulus results from a sufficient number of repetitions to show that intrasubject variance is small compared with intersubject variance.

The parameters characterizing the subject's sensitivity to various sweeteners do not rank in the same order from subject to subject; they do not vary in concordance. Moreover, some functions of intensity vs concentration are not linear even in log-log coordinates and some even present notches (Faurion 1982; Faurion et al. 1980). These characteristics depend both on the stimulus and the subject. This indicates that several mechanisms cooperate in signalling one stimulus and hints at differences in receptor equipment among subjects. The same reasoning as for interspecies differences leads us to assume the existence of one or other member of the following alternative:

1. There is one different receptor site for every subject, since all subjects exhibit different sensitivity profiles to the same set of sweeteners. This hypothesis is not realistic on a chemical basis because of the great variety of structures included in the experiment. A wide spectrum of sensitivities is not possibly accommodated by one and the same selective acceptor mechanism.
2. There is a different combination of several receptor sites for each individual which can give many different sensitivity spectra on a qualitative as well as a quantitative basis. On a qualitative basis every subject may not possess all receptors; on a quantitative basis subjects can possess various amounts of the same receptor site types.

More recently, Ferrell (1984), studying the relative stimulatory effectiveness of sugars on C.T. bundles in puppies and in three adult human subjects, showed a differential ranking for each individual.

3.2.3 Markers of Different Sensitivities

Thresholds and intensity functions for thaumatin, chloroform, D-leucine and sucrose differ the most across subjects, whereas the thresholds are more similar for stevioside and thaumatin and for saccharin and cyclamate. Fructose is not equivalent to sucrose, although it is not very different; neither is monellin equivalent to thaumatin etc. This can be laboriously seen using paired comparisons and more easily visualized in a spatial representation of discrepancies of subjects' sensitivity profiles obtained using multidimensional analysis.

The factor analysis by Benzecri et al. (1973) is a multidimensional analysis. It can be fed the original rectangular data matrix (subjects × stimuli) in order to extract the dimensions of a space containing the data variance (the greater the variance, the more numerous the dimensions) and to describe the co-variance between subjects as well as between stimuli in a vector space (where the great distances represent non-co-variance and short distances co-variance). Since it is a vector space, highest distances are represented on orthogonal directions. Stimuli associated with orthogonal vectors are independent; that is, their profiles across subjects do not vary in concordance. In other words, if subject A is highly sensitive to sweeteners a and b, it does not mean necessarily that subject B, who is highly sensitive to sweetener a, as well, will also be highly sensitive to sweetener b. Conversely, positive correlation between given profiles is indicated by proximity in the space generated by factor analysis.

D-leucine, chloroform, thaumatin and sucrose are supported by vectors which are orthogonal to one another, whereas saccharin and cyclamate, for example, are close to one another, as are stevioside and thaumatin etc. (Faurion et al. 1980). On the basis of such results, we suggest that sucrose, thaumatin, chloroform and D-leucine interact mostly with independent receptor sites although they are not specific to these sites and can interact with others, but quantitatively less.

3.2.4 Number of Receptor Mechanisms

Since the variance of individual data for seven sweeteners at threshold level needs at least five dimensions to be represented, it means that at least five independent factors are responsible for this variance. Factors responsible for sensitivity to sweeteners must be related to chemoreception mechanisms at the interface level between external medium and internal medium. Hence, we have suggested that five independent receptor mechanisms, at least, (which might be receptor proteins) are needed to explain differential sweet taste sensitivities in the human (Faurion 1982).

The corollary proposition is that several sites will signal each stimulus, presenting various affinities for the considered sweetener. We can thus explain the aspect of a continuum which is found in the factor analysis representation.

3.2.5 Quality Similarity Judgments

Meanwhile, studies in man have been performed using qualitative psychophysics. Schiffman et al. (1979) analysed similarity judgments of 12 subjects to 17 sweet stimuli. Subjects were asked to quantify the similarity of all paired sweeteners. Multidimensional data analysis of such ratings yielded a three-dimensional space where the most similar sweeteners to all subjects are the closest and stimuli judged different by all subjects are distant from one another.

Interestingly, the taste space representation obtained by both methods – qualitative and quantitative psychophysics – show some common points. Both spatial representations of sweet taste appear as continua; thaumatin is highly different from sucrose both on a qualitative and quantitative basis, either extracting the interindividual differences or pooling (averaging) subject's results, as Schiffman does. It may be that both treatments reveal two aspects of the same global mechanism. We suggest that we possess a continuum of sensitivities instead of an unique sweet sensitivity and several receptors to signal them. These different receptors could also signal qualities which are perceived somewhat differently since the same fibres code for quality as well as for intensity in the pattern theory (cf. Part A).

If thaumatin is the best (?) marker of a receptor site distinct from a sucrose receptor site, then the perceived quality elicited by thaumatin can be different from sucrose sweet taste through slightly different coded patterns. However, thaumatin taste may not be as different from sucrose sweetness as bitter is different from sweet taste. Pooling subjects' data will cancel the indication of the existence of different relative amounts of several sites in each individual, but will extract the global differences of perceived qualities.

3.3 Differential Sensitivity Profiles Among Single Neural Elements

Only the geniculate ganglion (Boudreau and Alev, 1973; Boudreau et al. 1982) and the C.T. responsiveness have been tested with more than the four standards. Some authors tested several salts (Sato T. 1980; Erickson et al. 1980, among others), but very few used several sweet stimuli. Apart from Andersen et al. (1963), who tried several sugars on a few fibres in order to study the interaction between salt and sugar stimulation, only Noma et al. (1974), Frank (1973), Sato M. et al. (1975), Jakinovitch (1981) and Beidler and Tonosaki (1984) used several sweet stimuli. Pritchard and Scott used several amino acids (1982).

Noma et al. (1974) compared the effectiveness of four alpha and beta anomers of sugars and sucrose on five hamster C.T. fibres. Observation of the raw data (p. 258) indicate that the response profiles to sugars are different from one fibre to another, as our human sensitivity profiles are different from one subject to another. It was already known that response profiles of fibres differed as far as the four standards were concerned, but it had not been apparent that fibre sensitivity could vary without concordance within the range of sweet stimuli.

Other authors added saccharin to the set of four standard stimuli. Ogawa et al. (1968, 1969) showed that Na-saccharin could elicit responses on sucrose-sensitive fibres or on NaCl-sensitive fibres in the hamster. It is usually considered that Na-saccharin is sweet, salty and even bitter to the rat as shown by behavioural experiments (Morrisson and Jessup 1977). Accordingly, Sato M. et al. (1975) have shown that quinine-best, sucrose-best and salt-best responding fibres could respond to Na-saccharin in the monkey. But, what is to be understood if some sucrose-sensitive fibres do not respond to Na-saccharin – as already shown in the dog by Andersson et al. (1950)?

We have recorded intrapapillar fibres in the hamster responding to sucrose, but not saccharin and vice versa (Faurion 1982; Faurion et al. 1983). Many other examples of non-covariance can be found involving other sweet-tasting stimuli. Such recorded fibres innervate sensory cells which seem devoid of a saccharin-sensitive receptor site, though they respond to sucrose. Hence sucrose and saccharin do not interact with exactly the same receptor sites. This hypothesis is globally exemplified and argued by across-fibre correlation coefficients between sucrose and saccharin which do not approach 1, and hence, indicate a distinction between the fibre response pattern to sucrose and saccharin ($r = 0.61$ in Frank's results in hamster, $r = 0.81$ on the monkey C.T., Sato and Beidler 1975). However, saccharin being sucrose-like in the hamster (Nowlis et al. 1980) is supposed to interact with the so-called "sweet" receptor site. This, along with differential sensitivity profiles, is a strong argument to suggest that molecules tasting sweet to the human can interact with several receptor sites and not only one.

Two hypotheses can be proposed: (a) either there are two specific receptor sites, one for sucrose and another for saccharin, or (b) there are two or several non-specific receptor sites which exhibit preferential affinity, one for saccharin, the other for sucrose.

In the first hypothesis, sucrose and saccharin-responding fibres are connected to cells which globally possess both receptors, whereas the sucrose-sensitive fibres which do not respond to saccharin are connected to cells which all lack one of these receptors. In the second hypothesis various amounts of each type of poorly specific sites can account for relative responsiveness to either sweetener.

We do not favour the first hypothesis, having ourselves recorded single units in the hamster papillae which each exhibited a different spectrum of sensitivity to 23 sweeteners (Faurion 1982; Faurion and Mac Leod 1982; Faurion et al. 1982). This is in close correlation with the findings of Noma et al. (1974), and with those of Boudreau, who reported that each geniculate ganglion single unit exhibited its own different sensitivity spectrum to a variety of stimuli (Boudreau et al. 1975) These data are in accordance with the findings of Sato M. et al. (1975), who tested saccharin, dulcin, lead acetate and cyclamate on the monkey. Across-fibre correlation coefficients between sucrose and each of these sweeteners are, respectively, 0.81, 0.71, 0.72, 0.21, which suggests a continuous range of sensivities rather than a unique discrete sweetness sensitivity. Equivalent figures can be found in Frank's data (1973).

More recently, Beidler and Tonosaki (1984) have also shown differential responses of C.T. fibres to a set of sugars; Tonosaki and Funakoshi (1984b) found the same with intracellular recordings in the mouse.

Because of the high number of different spectra obtained using many sweeteners, it is unwise to suggest that each fibre exhibiting a different profile is connected to cells containing exclusively one highly selective type of receptor.

Conversely, several receptor sites of poor specificity are likely to determine a wide variety of fibre sensitivities. Using factor analysis to treat a matrix of responses of 28 units to 12 sweeteners or of 18 units to 19 sweeteners, among other stimuli, we have found at least six independent factors to explain the fibre response variance to sweet-tasting molecules. If this means that at least six receptor sites might be involved in sweetness chemoreception, we should find at least 2^6 or 64 different sensitivity profiles. This illustrates the case of an all-or-nothing affinity of each receptor for each ligand. If we assume that receptors are most probably not specific but present a quantitatively different affinity for each stimulus, the number of different sensitivity spectra can be far greater and indeed infinite. Since we found a continuum of fibre sensitivities, we favour this last hypothesis although there is no electrophysiological demonstration for it; we must test far more than 64 fibres to develop strong evidence and electrophysiologists might not be able to sample so many.

3.4 Cross-Adaptation

Cross-adaptation is another type of quantitative assessment which can be used in human psychophysics. The procedure usually consists of presenting two different stimuli in the following order: A B A. The first presentation is used to determine the intensity of A in normal conditions; after a long enough delay, stimulus B is applied as an adapting solution and A is retested just after B. The difference of estimated intensities to stimulation by A before and after B is believed to be a measure of the effect of stimulus B on the receptor sites which interact with stimulus A.

McBurney (1972) has already found a differential reduction of perceived intensity of sweet taste by having subjects test various sweeteners after sucrose adaptation. Although it was an indication that there might be several acceptors, it was not absolute proof since he had used only sucrose as an adapting solution. Hence, the differently adapted sweeteners could have had various affinities for one and the same "sweet receptor", and therefore various abilities to displace sucrose from the site, assuming, for example, that adaptation might result from competitive inhibition.

Schiffman et al. (1981) applied a cross-adaptation procedure to a set of sugars and artificial sweeteners tested for reciprocal cross-adaptation. The results show that Na-cyclamate is best adapted by glucose. Na-sacharin is best adapted by acesulfame-K, less by Ca-cyclamate and glucose, but not at all by aspartame. D-Tryptophane is best adapted by aspartame, a little less by glucose and not at all by acesulfame-K or Ca-cyclamate. On the contrary, cyclamate provokes an enhancement of D-tryptophane intensity.

That some sweeteners do cross-adapt unequally means that they probably do not interact in the same way with the same receptor sites. Among others, there is probably a receptor site which is shared by Na-saccharin and acesulfame-K preferentially, whereas another one is more sensitive to D-tryptophane and aspartame. Notice that all quoted sweeteners are adapted by glucose, indicating a relatively universal interaction of glucose with all considered receptor sites. The same can be said for other sugars used in the same work.

Adaptation levels are on the whole low (40% of intensity reduction) considering all stimuli. This low level suggests a small overlap of the sets of receptor sites binding both adapting and adapted stimuli. A lack of cross-adaptation suggests that the considered stimuli bind with different receptors independently. Yet, there is probably not as much independence as the 60% left since some cross-adaptation may exist which does not appear statistically significant.

More recently, Lawless and Stevens (1983) have shown different levels of cross-adaptation between sweeteners such as saccharin, aspartame, dihydrochalcone, and acesulfame-K. Interestingly, sucrose appears to cross-adapt nearly all other stimuli, at least, it is the more efficient. This is well in accordance with our data. In results of factor analysis, whether the data is psychophysical or electrophysiological, at threshold or at supra-threshold level, sucrose is always very near the centre of gravity of the cloud of points which is the location of the maximal co-variance with all other stimuli. This constant finding indicates a kind of universality of sucrose which might be due to the multiple possibilities of hydrogen bonding of such sugars (Faurion and Mac Leod 1982). As a matter of fact, these authors have shown that the areas potentially involved in intermolecular interactions between sweeteners and receptors are wide as far as the stimulus structure is concerned, except for sugars. On the contrary, in sugars reacting areas are small but very numerous. There are probably many ways for a sugar to interact with receptors and probably all different "sweet taste" receptors are compatible with all or nearly all sugars.

3.4.1 A Complex Model of the Receptor Mechanism?

Schiffman and co-workers found some surprising enhancements, as opposed to the expected adaptation, which are not easily explained. Although seldom reported, enhancement cannot be taken as an artefact. Three out of five subjects gave a significantly enhanced response to glycyrrhizic acid after pronase E application (Faurion et al. 1979). Should we suggest that a glycyrrhizic-sensitive receptor eliciting a quality other than sweet taste (i.e. participating in the elaboration of a neural pattern, the resulting quality of which is not sucrose-like) has been more inhibited than the co-called "sweet" receptor site, so, that a residual pattern eliciting a sweet sensation is unmasked?

Or should these enhancements be related to a complex mechanism of the receptor protein? Reports of synergy between monosodium glutamate and nucleotides (Yamaguchi 1967; Rifkin and Bartoshuk 1980) suggest an allosteric model as described by Monod et al. (1965). Although Mooser (1980) has recently shown that C.T. responses to NaCl can be fitted by Monod et al. equations, there is presently no evidence that chemoreceptors are allosteric. Nonetheless, Tateda (1967) had already suggested a multimolecular combination of stimulus with the receptor site to explain the electrophysiological response to glycine on rat C.T. He observed a synergistic effect between sucrose and glycine. He also observed, that the variation of glycine responses with concentration did not at all obey the same function as sucrose or fructose. Hence, glycine data could not fit Beidler's equation (1954, 1962), but formed a relation similar to Hill's (1913) which suggested the cooperation of at least four glycine molecules to the glycine-receptor complex.

Can all these observations be explained in terms of several independent receptors without allosterism? This is unclear as far as enhancements and synergies are concerned. They suggest a derepression or a transconformation of the receptor towards a structure of the receptor presenting a better affinity. This is usually referred to as the two-stage model receptor site.

Conversely, could an allosteric unique receptor give account of multiple sweetness sensitivities? No, because of interindividual differences.

3.5 *Action of Inhibitors*

Some researchers tested the effects of inhibitors known to depress response to the sweet taste of sucrose and other sweeteners. As in the cross-adaptation experiment of McBurney (1972), the magnitude of the response reduction obtained after inhibition varied with the stimulus used even if the stimuli had been applied in such concentrations that the responses obtained before the action of the inhibitor were of equal amplitude.

Yamamoto and Kawamura (1971) did the same experiment on C.T. responses of the rat to sucrose, fructose, glucose, glycine, alanine and lead acetate. Application of Cu^{++} on the tongue produced inhibition of the responses which was equal in amplitude for all sweeteners used, whereas Zn^{++} provoked a differential inhibition. For example, sucrose, glucose, fructose and saccharin were more inhibited than L-alanine. Glycine was less inhibited than L-alanine; $PbCl_2$ was even less inhibited.

Moreover the authors showed that Zn^{++} inhibition was competitive, whereas Cu^{++} inhibition was non-competitive, suggesting that Zn^{++} could interact with the receptor site itself and Cu^{++} with more general properties of the membrane. Iwasaki and Sato (1984a) resumed this study in the mouse.

In the same kind of experiments, Hiji and Ito (1977) applied pronase E to the tongue of the rat while recording on the whole C.T. The responses to alanine and glycine were less depressed than the responses to glucose, fructose, sorbitol and Na-saccharin; moreover, looking at their figures, it seems that the response to fructose is not as fully suppressed as is the response to glucose, although they appeared identical before application of pronase E.

Two hypotheses must then be considered. (a) After it has been modified by an inhibitor, a unique receptor site presents an affinity order to sweeteners which is different from what it was before, and the previously equilibrated concentrations no longer give responses of equal amplitude. (b) Several acceptors interact with the tested stimuli, perhaps only some of them are modified by the inhibitor.

Experiments have been undertaken in our laboratory (Faurion and Mac Leod 1982; Faurion et al. 1979) to test these hypotheses. Two experiments have been conducted on man, one with gymnemic acid (G.A.) the other with pronase E.

Magnitude estimates of four concentrations of 20 sweeteners were obtained from five subjects before and after application of a controlled dose of pronase E on the subject's tongue. Each concentration was tested about 25 times by each subject so that individual results could be expressed within a relatively small confidence interval. It appears that the individual profiles of intensity change for each of the 20 sweeteners after pronase E do not co-vary either across subjects or across sweeteners. It is not possible to state that sweetener A is globally more depressed than sweetener B by pronase E; it all depends on the subject.

This result indicates that all sweet stimuli tested must interact with several mechanisms which are unequally distributed among subjects. Moreover, some sweeteners which did co-vary before the application of pronase E did not co-vary any more after pronase E application, as if one or several receptors more sensitive than others to these sweeteners had been selectively depressed. For example, sucrose, sorbitol, saccharin, Na-cyclamate, methyl α D-glycopyranoside (MGL), acesulfame-K and fructose sensitivities which did co-vary among these five subjects before action of pronase E were suddenly non-covariant after pronase E application. Only sucrose and sorbitol, on the one hand, and fructose and MGL, on the other, still co-vary. It is not certain that we would have obtained exactly the same results with other subjects, since five subjects are actually a very small sample; but, it is still interesting to demonstrate that in a given subpopulation, sucrose and sorbitol have (a) receptor(s) in common, fructose and MGL have (an) other receptor(s) in common and that yet (an)other receptor(s) can accept all of them non-selectively (Faurion 1982).

Results of inhibition of the sweet taste of seven stimuli in nine subjects by G.A. gave the same kind of differential inhibition. As in all previous experiments, thaumatin, chloroform and D-leucine were the most different from sucrose.

Hellekant and Roberts (1983) showed such quantitative discrepancies on 12 single units recorded in the hamster C.T. before and after application of G.A.; the difference of response magnitude, and the direction of this difference, to xylitol, acesulfame-K, sucrose, glucose, fructose and saccharin, depend on the sweetener and on the fibre.

Do these several receptors actually cooperate to perception of sweet or other tastes? Indeed, subjects were asked and trained to estimate the intensity of sweet taste only and the dose of pronase E had been chosen so that the sweetness reduction was small for all sweeteners for all subjects; hence, an estimation of the intensity of the residual sweetness was still possible after pronase E application. Therefore the different receptors disclosed by pronase E in this experiment should actually all cooperate to sweet taste perception.

Should these receptors cooperate to perception of other tastes as well?

No argument for or against this hypothesis can be extracted from the preceding data. The procedure of sweet taste intensity measurements excludes the possibility of adding representatives of other tastes to the stimuli set.

But more recently, Jakinovich (1983) tried to inhibit the response of the gerbil whole C.T. to sucrose by 4,6-dichloro, 4,6-dideoxy, α D-galacto-pyranoside, a compound sweet to the human. He showed that sucrose responses were inhibited competitively, whereas saccharin responses were inhibited non-competitively. It can be argued that the sweet inhibitor interacts with the same site as sucrose and that saccharin interacts with a least one other one. But, it is well known that saccharin is bitter as well as sweet. It might well be that the part of the response which leads to a non-competitive inhibition could be due to a so-called "bitter" receptor site.

Therefore, we must now gather evidence to try to answer the question whether the several receptor sites which cooperate to the sensation elicited by sweet-tasting molecules cooperate to the chemoreception of non-sweet-tasting molecules, for example bitter taste. This point will be developed in Section 6.

Mode of Action of Inhibitors. In these experiments the precise site of action of the inhibitor is not actually known. Yamamoto and Kawamura suggested that Zn^{++} could interact with the receptor site itself; the same is suggested for the sweet dichloro dideoxy galactopyranoside. But no author has reached any conclusion about the action site of gymnemic acid (Meiselman & Halpern 1970). Warren and Pfaffman (1959) found that saccharin and glucose were inhibited in the same way by G.A. Should their measurements be comparable to those of Jakinovich, it would mean that G.A. and dichloro dideoxy galactopyranoside do not interact in the same way with the sweet acceptor mechanisms. Glaser et al. (1984) showed that G.A. does not suppress the taste of sucrose in non-human primates. We can therefore deduce that G.A. inhibitory effect is probably structurally related to the acceptor. Another inhibitor extracted from a plant, ziziphin (Kennedy and Halpern 1980a,b; Smith and Halpern 1983), might interact in a similar way with the receptor.

Pronase E is most probably acting on the chemoreceptor proteins by proteolysis unrelated to the receptor site structure (Nomoto et al. 1960; Narahashi 1970; Kitada 1984).

3.6 Cooperation of Receptors Involved in Qualitatively Different Perceptions of Sweet Taste

Many converging arguments have been gathered in this review to defend the hypothesis of the existence of several receptor sites cooperatively contributing to the sensitivity to stimuli which are sweet to man. At least, enough arguments invalidate the hypothesis of the existence of only one receptor site sensitive to sweet stimuli. The question then arises whether these cooperating sites can be involved in bitter or other tastes?

This question has much to do with the sempiternal question whether taste is a continuum or a combination of four distinct qualities. It must be emphasized that taste quality is a difficult concept to handle since it does not match modern electrophysiological or biochemical data. It has been first used by psychologists for human perception and physiologists have indulged in using it in working on neuron sensitivity or responsiveness. The situation is even worse when trying to match receptor sites with the qualitative concept. It makes imprecise, misleading sense.

In this respect, it is useful to redefine the hypothesis of a continuum for taste at three independent levels: receptor site level, neuron sensitivity level (including the whole sensory chain), and perceptual level. Because sensory cells already contain randomly distributed receptors to all qualities, the nervous system could be responsible for either convergence or even more divergence of information.

Four distinct mechanisms,[7] as well as a greater number of receptor sites could give rise to a continuum of neuron sensitivities. Conversely, any number of sites could define a set of subcontinua centred on each of the four qualities, as well as a unique continuum, for neuron sensitivities (Pfaffmann 1974). Should we possess either a set of subcontinua or a unique continuum of neuron sensitivities to the whole range of taste stimuli, CNS decoding might translate it into either a continuum of perceptions or four distinct qualities. Figure 2 illustrates all combined possibilities. Let us sum up the trends that have been indicated by literature on data at all three levels.

---

[7] As a matter of fact, four independent mechanisms are already known for salt, acid, bitter or sweet taste. As salt and acid may depend on other types of mechanisms such as ionophores for Na^+ and Li^+, or a pH-related membrane property for acid taste, we prefer using the concept of mechanism rather than of chemoreceptor

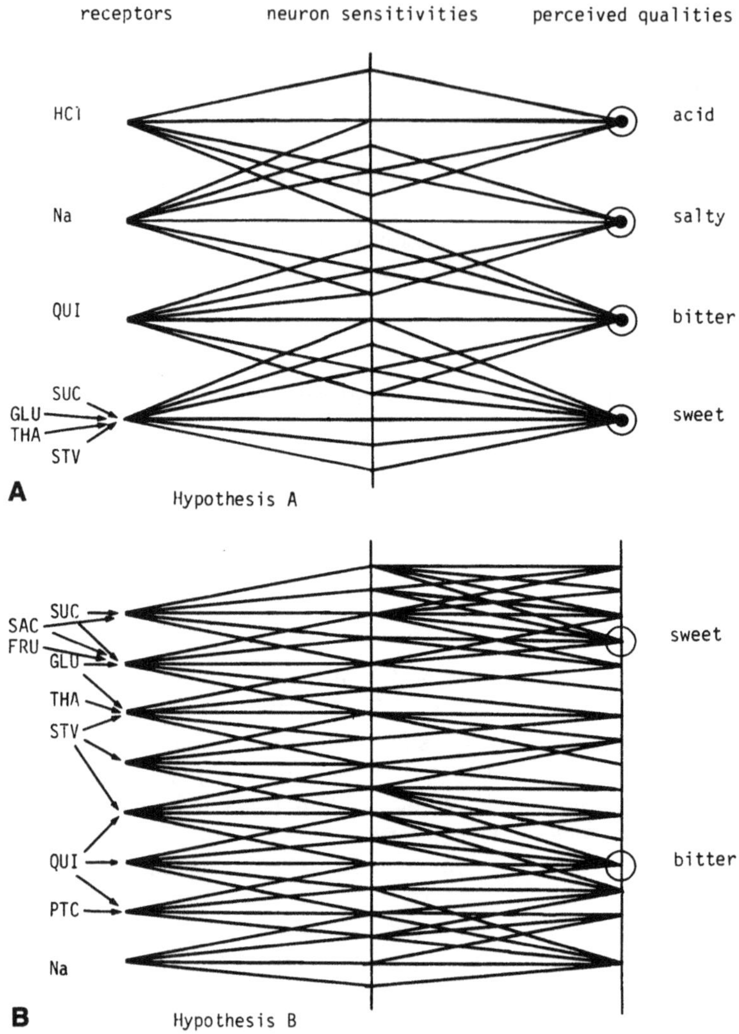

Fig. 2A, B. Graphic representation of two possible taste coding organizations from receptor level to perception level. Sets of marks represent patterns which convergingly or divergingly project onto the next level. **A** Four receptor sites, a continuum of sensitivities, four discrete qualities. **B** More than four receptor sites, a continuum of neural sensitivities, a continuum of perceived qualities with four reference marks privileged by semantic descriptors. Two other possibilities, which are not represented here could be logically envisioned i.e. (**A**) four receptor sites diverging onto a sensitivity continuum and a quality continuum; (**B**) more than four receptor sites diverging onto a sensitivity continuum that would converge onto a set of four discrete qualities. Because of recent electrophysiological data, we submit that hypothesis B (similar to Pfaffmann's second one), best represents trends of present knowledge on taste physiology

At the perceptual level, do we possess only four qualitatively distinct perceptions? We have seen that four qualities do not account for alkaline, metallic, monosodium glutamate or glycyrrhizic acid tastes. Henning (1916) proposed the expression of a continuous range of taste qualities (*kontinuierliche Reihe*). Erickson pointed to the epistemological error in quoting Henning as a reference for four basic tastes (Erickson and Covey 1980; Erickson 1983; Faurion 1983). Schiffman and Erickson (1980), Schiffman et al. (1980), and Woolston and Erickson (1979) still argue in favour of this expression. The fact that taste is neither synthetic nor analytic (Erickson and Covey 1980; O'Mahony et al. 1983) destroys the idea that four distinct qualities should combine to produce all taste perceptions. Arguments for a continuum of perceptions invalidate the idea of four basic qualities without being able to prove the contrary.

Concerning the sensitivity level, it seems that we should be able to gain a better knowledge of the situation through experimental investigation on single units using a wide set of stimuli. Erickson et al. (1980) used several salts which appeared interspersed in a sensitivity continuum in between acid, sweet and bitter taste. Recent experiments (Faurion et al. 1983) on intrapapillar single units argue in favour of a unique continuum of sensitivities for taste, since a given fibre may respond or not to any of the 23 tested sweeteners, whatever its response to the four standards might be.

But neither the knowledge of the structure of the sensitivity organization at neural level, nor of perceptual space is of any help in understanding the organization at the chemoreceptor level.

Arguments for and against the Involvement of Receptors in the Perception of Several Qualities. With the diabetogenic compound alloxan, Zawalich (1973) was able to depress rat C.T. responses to sucrose, mannose, arabinose, 2-deoxy D-glucose and xylose, but he could not depress the responses to the other three standards. Noma and Hiji (1972) could depress rat C.T. responses to sucrose only among the four standards with thiol re-agents, such as p-chloromercuribenzoate (pCMB) and N-ethylmaleimide (NEM), thus indicating that SH groups are related to the function or the structure of the "sweet" chemoreceptor. Even though, Hidaka (1970) demonstrated the depression of electrophysiological responses to NaCl and sucrose with NEM in the carp, one can argue that fish and mammal receptors may be very different in structure. Imoto et al. (1981) have shown that another thiol re-agent, S-mercuric N-dansyl cystein, depresses only sucrose responses in the rat C.T. The extract of Ziziphus jujuba depresses intensity judgments to sweet taste only in man (Meiselman et al. 1976).

Concerning pronase E, Yoshii et al. (1981) have shown that the enzyme has no action on frog glossopharyngeal responses to quinine but to sucrose only, whereas Hiji (1975) stated that "it hardly affected other tastes".

Most discrepancies emerge among author's reports regarding gymnemic acid. McBurney and Gent (1978) have shown that G.A. reduces the sweet taste of methyl D-mannopyranoside, whereas the bitter taste of this compound, which is known to be ambivalent (Birch and Lindley 1973) is not reduced. Riskey et al. (1982), too reported that G.A. has no action on other qualities. Mehren and Church (1976) worked on the pigmy goat, which is peculiar in the way it naturally prefers quinine and rejects sucrose. G.A. added to each stimulus converts the mild rejection of sucrose into strong preference and the preference to quinine into a moderate rejection, if quinine concentration is kept low. This is quite puzzling but, of course, goat receptors may not be similar to ours. To go even further from the human, Kennedy et al. (1975) have shown that behavioural responses to sucrose and NaCl in the house fly are affected by G.A. as if it were a narcotic. Hellekant et al. (1974) and Hellekant (1976) have shown that whole C.T. responses to sucrose are not depressed either in the rat, guinea pig, rabbit, pig or Cercopithecus monkey. In the dog, the effect depends on the individual, and responses to NaCl are also depressed but for a shorter time. In the rabbit, it seems on the figure presented by the authors, that the responses to NaCl are also somewhat depressed.

In 1959, Warren and Pfaffmann reported that G.A. interacted with bitter and sweet tasting compounds. Yacksan (1966, 1969) observed reductions to NaCl, HCl, and quinine hydrochloride. Bartoshuk et al. (1969) found a non-significant reduction for low concentrations of quinine and argued that the previously shown bitter taste reduction might have come from an adaptation due to the bitterness of the G.A. extract.

None of these data actually confirm that receptors involved in sweet taste would be involved in other tastes and vice versa. Nonetheless, we have found an interesting clue in a paper from Meiselman and Halpern (1970). They had subjects estimate the intensity of sweetness, bitterness, sourness and saltiness, independently, of sucrose, DL alanine and glycine solutions, before and after action of G.A. Sweetness of all three compounds was reduced. Bitterness was enhanced for sucrose and DL alanine, but not glycine. Glycine acts as a control to show that bitter taste was not elicited by G.A.; neither glycine nor alanine have ever been reported in the literature as tasting bitter.

The enhancement of bitter taste suggests that sucrose, D and/or L alanine interact with bitterness-sensitive receptors as well as with sweetness-sensitive receptors. The result of this interaction may be usually not perceived, being a small portion of the information which is overlooked with respect to sweet taste. However, it may appear when sweet taste is reduced.

Bartoshuk (1979) showed that the bitter taste of saccharin is related to the genetic ability to taste the bitter substance 6-n propylthiouracil (a substance dividing human population between so-called tasters and non-tasters). It seems that bitter taste and sweet taste elicited at low concentrations by Na-saccharin is more intense to tasters than to non-tasters. Could it be that saccharinate ions interact with receptors, among them the more sensitive 6-n propylthiouracil receptor? Should that be the case, the quality of the pattern could slightly change between tasters and non-tasters.

Moreover, Gent and Bartoshuk (1983) have shown that the sweetness of sucrose, neohesperidin dihydrochalcone (DHC) and saccharin is also related to the genetic ability to taste the bitter taste of 6-n propylthiouracil. In this case, the pattern is still interpreted as sweet by the central nervous system, probably because its overall form "looks" sucrose-like as originating mainly form receptors which accept sweet tastants as a majority rather than bitter tastants. However, it is an indication that these molecules may interfere with several receptors among which one at least may contribute to another quality. When a bitter tastant is tested, this peculiar receptor is triggered as well as other bitterness-sensitive receptors, thus

eliciting an overall "bitter pattern", however, some sweetness-sensitive receptors may have been triggered (the effect of which is overcome if they are in the minority).
This hypothesis is a rather satisfactory way to explain that methyl alpha D-mannopyranoside (MMP) is both sweet and bitter. The elicited pattern is a set which includes (union of sets) a bitter-like and a sweet-like pattern. By chance, neither overcomes the other.
When dealing with such reasoning, we mut be sure that the tastant is actually *pure* in solution – which is unreachable. Imagine a sucrose solution in which one out of ten thousand parts (1/10000) is an impurity of, for example a bitter-tasting nature – and more than 10000 times more efficient than sucrose; this is actually possible since we use grams per liter for sucrose and mere fractions of milligrams per liter for many tastants, especially bitter ones. We cannot check such purities using HPLC. Only varying conditions of preparations and cross-indications can give us a hint.
However, and because we have more than only one example (sucrose, saccharin, DHC, MMP), I think it is presently sensible to suggest that sweet- and bitter-tasting molecules, for example, might partially interact with common receptor sites, i.e. a small proportion of the receptors which bind sweet-tasting ligands might also bind bitter-tasting ligands. A recent experiment (Jakinovich 1982) which shows that methyl glycosides are avoided by animals taught to avoid sucrose (in this case gerbils), as well as by animals taught to avoid quinine-HCl, reinforces this hypothesis.

3.7 Protein Fractions and Binding Experiments

Several attempts have been made to extract the so-called sweetness receptor protein, that is the chemoreceptor that binds stimuli which are perceived as sweet by man. Dastoli and Price's (1966) first extraction was not conducted in parallel with tongue epithelium devoid of taste buds and Koyama and Kurihara (1971) easily showed that the sugar-binding protein extracted by the former authors could in fact be extracted from tongue epithelium devoid of taste buds. This is a demonstration that sugars can bind to tongue proteins which are not associated to any chemoreception function and this binding is called non-specific binding.
Subsequent workers always took care to test specific and non-specific bindings comparatively on protein fractions originating from epithelium containing taste buds and on epithelium devoid of taste buds (Hiji and Sato 1973).
Of course, there is no proof that specific binding, i.e. the binding which is found only with taste-bud-containing fractions is actually related to actual chemoreceptors. As Cagan and Morris (1979) wrote, "no convincing demonstration of the isolation of taste receptor molecules has yet appeared".
Kadowaki et al. (1979) tested comparatively, using electrophoresis, the protein composition of homogenates of either fungiform or circumvallate papillae of the dog, monkey and man. In the dog, one or two bands were specifically found only in the fungiform papillae of four individuals out of six. One band existed only in the circumvallate papillae of three out of six dogs. In the human, one to four proteins, depending on the subject, were found only in the fungiform papillae; one or two proteins were specifically found only in circumvallate papillae of four

out of five subjects, whereas one to three proteins existed both in circumvallate and fungiform papillae. As in vivo circumvallate and fungiform papillae are known to exhibit different sensitivities, and as we have gathered arguments in favour of a different chemoreceptive equipment explaining interindividual sensitivity differences in the human, it is tempting to think that these protein bands could be related to chemoreceptors.

There are nonetheless other arguments in favour of the hypothesis that in vitro specific binding could be related to the chemoreception function in vivo.

Hiji et al. (1969, 1971) and Hiji and Sato (1973) reported the binding of a protein fraction with several sugars which paralleled the magnitude order of electrophysiological responses in the rat, whereas they could obtain no binding with rats – which had been treated with colchicine (devoid of taste buds) or with cats – which are known to be poorly sensitive to sweet stimuli.

Sato M. et al. (1977a,b) obtained binding between monkey tongue extracts and sucrose or aspartame, but very poor binding between aspartame and rat tongue extracts. This parallels an electrophysiological finding which shows that rat C.T. responses to aspartame are very low.

Cagan (1971) showed that bovine taste papillae suspensions could bind more ^{14}C-labeled sucrose than suspensions deriving from filiform papillae which are devoid of taste buds. Moreover, lactose elicits little preference behaviour as opposed to sucrose fructose and glucose and accordingly binds much less than the latter compounds.

No author reported that purified fractions containing only one protein bind to sugars. Conversely, they complained of the very low affinity of sugars for receptors; Kd values range between 10^{-1} and $10^{-3}\,M$ (Cagan 1971). Authors usually chose to bind sweet stimuli to homogenates or non-purified protein fractions. Moreover, negative findings are reported. Ostretsova et al. (1975) failed to show any binding with glucose. Hiji and Sato (1972) reported a very small amount of protein binding to fructose in rat tongue extracts. This means that glucose or fructose affinity for the chemoreceptor is not much higher than their affinity for any membranous protein which has no chemoreceptive function (non-specific binding).

Cagan and Boyle (1984) showed that the binding of seven chemostimulatory aminoacids to purified fractions obtained from catfish taste tissue "corresponds reasonably well with earlier binding data obtained with less pure sedimentable fractions". The binding is more or less correlated to the efficiency of stimuli as measured behaviourally. Although it is not possible to extract pure receptors, most of all because of the non-specific nature of the stimulus-acceptor relationship, it is fairly sensible to consider binding measurements as representative of this relationship. In these conditions, relative measurements are necessary and cross-binding of several fractions with several stimuli is a promising strategy.

Cagan and Morris (1979) have developed binding experiments on cow and human tongue homogenates with monellin which is about 10^4 to 10^5 times more effective than sugars in eliciting a sweet sensation in the human. They found that glucose and lactose could displace the complexed monellin by 37% and 32%, respectively. This could mean that glucose and lactose bind to about 40% of the acceptor sites sensitive to monellin.

Whether the 60% remaining should concern any specific binding, i.e. a binding leading to sweet sensation, is not known, but is possible. Whether it concerns other chemoreceptor proteins or only other acceptor sites belonging to one and the same receptor molecule is not known either. On the contrary, these results could mean that glucose and lactose cannot displace all the bound monellin because of their lower affinity for the same receptor site. But, with extracts obtained from human tongues, the same authors have found that sucrose, lactose, Na-saccharin, neohesperidin dihydrochalcone, Ca-cyclamate and aspartame could displace 20% to 40% of the bound monellin, whereas stevioside had no effect; however, stevioside is usually considered as sweet for the human. Could stevioside bind to a receptor site other than the monellin-sensitive receptor site(s)? Shimazaki et al. (1981) showed that the binding of saccharin to a 40% – 60% ammonium sulfate fraction from rat tongue epithelia was inhibited by 35% by 1 *M* glucose, whereas it was fully inhibited by 10 m*M* saccharin. Shimazaki and Sato M. (1981) tested the binding of several sweeteners to monkey tongue extracts. They found, and this must be emphasized, that glucose and saccharin inhibit specific and non-specific binding to a lesser extent, whereas aspartame, stevioside, and thaumatin do not inhibit non-specific binding of either saccharin or aspartame. Moreover, they found that the specific binding of saccharin is inhibited by aspartame and saccharin (among others) but not by stevioside, whereas the binding of aspartame is inhibited by saccharin, aspartame and stevioside.

These results, which are coherent among independent authors, mean that saccharin, aspartame and stevioside can bind to a molecule, let us call it A, and that saccharin and aspartame can bind to another molecule, let us call it B, to which stevioside cannot bind. Both these molecules are specifically found in the fraction containing taste buds. Whether they both participate in chemoreceptive function cannot be proven by any biochemical finding. Nonetheless, it is rather impressive to find psychophysical and electrophysiological findings in favour of the existence of several receptor sites and, independently, in vitro evidence of the existence of several molecules which are specifically found in taste buds, but not in ordinary epithelium and which can differentially bind to sweeteners.

It must be emphasized that saccharin and glucose act as universal stimuli, being able to displace saccharin or glucose equally well on specific or non-specific complexes. This finding is consistent with cross-adaptation results. This is in accordance with the great difficulty the biochemist has in finding specific binding with these compounds; the chemoreception system, however, does not care that there is non-specific binding. As opposed to the biochemist, the brain does not see the non-specific binding, since it is not taken into account by sensory cells. On the contrary, what is of special interest to the chemoreception system is that chemoreceptors can be of low specificity so that a great number of different sweet molecules may be accepted by the same system which may hence be constituted by a small number of distinct receptor molecules. The corollary is necessarily a low affinity, hence, the little difference between affinity for chemoreceptors localized on receptor cells and for any other membranous proteins.

It remains very difficult to prove that chemoreceptors studied in vitro are actually gustatory; moreover, no biochemical data can presently argue for their participation in the perception of one or several qualities.

A new approach, involving monoclonal antibodies (Golstein and Cagan 1982) is very promising. It is the only way to select pure entities from a medium which cannot be purified because it contains non-selective and non-specific receptors, and, further it checks how they react comparatively with this medium. Up to now, only preliminary work has been reported and no cross-reaction results have been published.

3.8 Conclusion

There is probably no animal receptor system which is representative of the human system. Physiological studies, presently nearly all restricted to the chorda tympani, probably do not sample the whole range of taste physiology. Nonetheless, physiological experiments bring us some knowledge about receptor mechanisms and quality discrimination.

Receptor Level. The examination of a great number of sensitivity profiles obtained in response to stimulation by a collection of sweet stimuli leads to the hypothesis that chemoreception most probably depends on a small number (≤ 10) of weakly specific independent receptor sites, the combination of which allows a high number of different sensitivity spectra.

One should ask, What is their degree of specificity? In single-unit recordings, one can look for a lack of response which can be interpreted in terms of a lack of a given receptor site. For example, single units which do not respond to saccharin, although they are sensitive to sucrose, probably lack a saccharin-sensitive receptor site. Hence, there is some specificity.

If we are not any more interested in single units but in the whole set of them, i.e. whole chorda tympani recordings, it is well probable that we shall not find lack of responses any more. Other cells most probably possess one or several receptors which are sensitive to sucrose and to saccharin so that related patterns are analogous enough to both be interpreted as sweet. One can then ask, Should there be two specific receptor sites included in the same cells or one (or several) non-specific receptor sites which can bind both sucrose and saccharin? Both hypotheses are not exclusive. Saccharin binding is in vitro displaced by sucrose and vice versa, hence, there must be some rather weakly specific receptors which can bind both ligands.

Data obtained with wide sets of stimuli show that taste chemoreception deals with a multidimensional system (depending on several independent factors): one receptor accepts ligands A, B, C, D; one accepts B, D, E, F, etc.

This exemplifies that data variance is sometimes information, not mere technical uncertainty. Data should not be reduced to a monodimensional argument if they possibly represent a multidimensional reality. Looking, now, for a specific relation between ligand and receptor would be as poorly heuristic as looking for no specificity at all (a receptor accepting all sweet-tasting ligands).

Quality Discrimination. What does quality emerge from? A set of receptors gives rise to a given pattern which is recognized and identified. If, by using n stimuli, we experimentally find n different spectra (representatives of whole patterns), it means that neuron sensitivity is to be expressed in terms of n patterns, not only four.

Is there a reduction into four qualities at higher levels? Two similar patterns, emerging from an interaction with a certain proportion of identical receptors, will be interpreted in terms of similar qualities. What is the degree of difference after which we decide that qualities are different? It could well depend on the "note" on which our attention focuses, that is supposing we are even able to get a subset of a pattern to resemble another one. Why so?

Semantics enables us to use several descriptors (sweet, bitter etc.) which act as reference marks. It is easier to make a judgment with respect to a reference than without it. We can, hence, privilege some marks within a continuous range of qualities by using semantic descriptors. The pattern is then not only recognized but *named,* thus stressed as a reference. Neighbouring points (qualities) will be qualified by reference to the nearest semantically designated quality, hence, drawn towards it.

Qualities which lie far apart from any reference will just be qualified as "complex" in the sense of complicated, since they cannot be easily reduced by simple analogy to any reference. They nonetheless do exist and constitute the taste continuum. But we only have an accurate knowledge of what we can name. What does glutamate taste like to the occidental man? What does glycyrrhizic acid taste like? (It has no odour; we have an experimental argument for that.) If we become trained to recognize them both, we shall soon be able to describe a given tastant as glutamate-like or glycyrrhizic-like; here are two newly generated qualities which merely lack semantic descriptors.

The quality concept is more semantics than sensory physiology. We are perfectly able to categorize taste qualities into four basic tastes if we are asked to, and in doing so, we accept quite a large degree of difference in the process of analogy recognition. Yet it does not mean that the CNS actually processes things so. We can program a computer to work as well as the human brain to execute some function; we shall not deduce from that that human brain process is the same as a computer process!

In fact, the central nervous system can recognize quite a lot of patterns up to memory saturation and we could develop as many semantic descriptors as we should like to, if it were of any use. It is not necessary to imagine a process of reduction of information into a small number of categories for giving account of the cognitive process of the perception of quality. A given stimulus is represented by a given pattern, i.e. a form. A form is here understood in the sense of Gestalt theory (Köhler 1930). If this form is memorized, it acquires the meaning of the contextual events relative to its memorization. Any gustatory sensation is a different form, a different pattern, even in taste, which is probably less rich than olfaction; there are far more than only four distinct sensations. There is no necessity for the CNS to reduce the information by a classification into categories before evaluating a sensation. According to Erickson (1983, 1984), the pattern *is* the message at any level of the CNS. If a proportion of firing fibres change, the form is enough different to elicit a different sensation. It is possible, as a subject, either to focus on the similar part of two distinct patterns and state, for example, aspartame is sweet or, to focus on dissimilarities between these two patterns and state aspartame is not exactly like sucrose.

If DHC is not salt-, sucrose-, HCl- or quinine-like in the gerbil, it means that gerbil cortex recognizes it as DHC-like and not as sucrose-like. When a human states that DHC is sweet, he voluntarily makes the analogy with sucrose, which it suggests, thus biasing the expression of the whole sensory information.

Indeed, some of our subjects do not easily agree that thaumatin is sweet. It is more glycyrrhizic acid-like to them, that is licorice-like. However, other subjects easily accept "sweet" as a label for glycyrrhizic acid. Different persons obviously do not experience the same sensations for these stimuli.
It is however always possible to make a categorization operation on a continuum. Anybody can classify a colour continuum into 3, 4 or 5, etc. categories. This highly integrative operation has nothing to do with the process of cognition related to colour perception. It is not because you first classified turquoise in the range of blues that you will be aware it is turquoise!
In fact, form recognition is a much more powerful device than information reduction in order to discriminate many different stimuli. If you need to discriminate among gustatory stimuli, you are able to; if you do not care about discrimination, you will blur this information – this does not mean it is not available at any level in the CNS.

Acknowledgments. I am indebted to Drs. Mac Leod and Pager for reading the manuscript and to Dr. Trotier for reading and encouraging discussion. Special thanks are due to Dr. Lindley for correcting Part B and to Dr. Erickson for correcting and commenting on Part A.

4 References

Andersen HT, Funakoshi M, Zotterman Y (1963) Electrophysiological responses to sugar and their depression by salt. In: Zotterman Y (ed) Olfaction and taste I. Pergamon, Oxford, pp 177–192

Andersson B, Landgren S, Olsson L, Zotterman Y (1950) The sweet taste fibres of the dog. Acta Physiol Scand 21:105–119

Azuma S, Yamamoto T, Kawamura Y (1984) Studies on gustatory responses of amygdaloid neurons in rats. Exp Brain Res 56:12–22

Bartoshuk LM (1975) Taste mixtures: is mixture suppression related to compression? Physiol Behav 14:643–649

Bartoshuk LM (1979) Bitter taste of saccharin related to the genetic ability to taste the bitter substance 6-n-propylthiouracil. Science 205:934–935

Bartoshuk LM, Cleveland CT (1977) Mixtures of substances with similar tastes. Sensory Processes 1:177–186

Bartoshuk LM, McBurney DH, Pfaffman C (1964) Taste of sodium chloride solutions after adaptation to sodium chloride: implications for the "water taste". Science 143:967–968

Bartoshuk LM, Dateo GP, Vandenbelt DJ, Buttrick RL, Long L Jr (1969) Effects of Gymnema sylvestre and Synsepalum dulcificum on taste in man. In: Pfaffman C (Ed) Olfaction and taste III. Rockefeller University Press, New York, pp 436–444

Bartoshuk LM, Rennert K, Rodin J, Stevens JC (1982) Effects of temperature on the perceived sweetness of sucrose. Physiol Behav 28:905–910

Bartoshuk LM, Desnoyers S, O'Brien M, Gent JF, Catalanotto FC (1985) Taste stimulation of localized tongue areas: the Q-tip test. Proceedings of the 7th annual meeting of the association for chemoreception sciences, Sarasota, abstract no 14

Bealer SL, Smith DV (1975) Multiple sensitivity to chemical stimuli in single human taste papillae. Physiol Behav 14:795–799

Beckstead RM, Norgren R (1979) An autoradiographic examination of the central distribution of the trigeminal, facial, glossopharyngeal, and vagal nerves in the monkey. J Comp Neurol 184 (3):455–472

Beidler LM (1953) Properties of chemoreceptors of tongue of rat. J Neurophysiol 16:595–607

Beidler LM (1954) A theory of taste stimulation. J Gen Physiol 38:133–139

Beidler LM (1962) Taste receptor stimulation. In: Butler JAV, Huxley HE, Zirkel RE (eds) Progress in biophysics and biophysical chemistry XII. Pergamon, New York, pp 109–151

Beidler LM, Tonosaki K (1984) Multiple receptors sites. Proceedings of the eighth international symposium on olfaction and taste, Melbourne, 1983. Chem Senses 8 (3):244

Beidler LM, Fishman IY, Hardiman CW (1955) Species differences in taste responses. Am J Physiol 181:235–239

Belitz HD, Chen W, Jugel H, Treleano R, Wieser H, Gasteiger J, Marsili M (1979) Sweet and bitter compounds: structure and taste relationship. In: Boudreau JC (ed) Food taste chemistry. American Chemical Society, Washington, pp 93–131

Benjamin RM (1963) Some thalamic and cortical mechanisms of taste. In: Zotterman Y (ed) Olfaction and taste I. Pergamon, pp 309–330

Benjamin RM, Akert K (1959) Cortical and thalamic areas involved in taste discrimination in the albino rat. J Comp Neurol 111:231–260

Benzecri JP et al. (1973) L'analyse des données, vol 1, vol 2. Dunod, Paris

Bernard RA, Nord SG (1971) A first order synaptic relay for taste fibers in the pontine brain stem of the cat. Brain Res 30:349–356

Biedenbach MA, Chan KY (1971) Tongue mechanoreceptors: comparison of afferent fibres in the lingual nerve and chorda tympani. Brain Res 35:584–588

Birch GG, Lindley MG (1973) Structural functions of taste in the sugar series: effects of aglycones on the sensory properties of simple glycoside structures. J Food Sci 38:665–667

Birch GG, Latymer Z, Holloway M (1980) Intensity/time relationships in sweetness: evidence for a queue hypothesis in taste chemoreception. Chem Senses 5 (1):63–78

Block CH, Schwartzbaum JS (1983) Ascending efferent projections of the gustatory parabrachial nuclei in the rabbit. Brain Res 259:1–9

Boudreau JC, Alev N (1973) Classification of chemoresponsive tongue units of the cat geniculate ganglion. Brain Res 54:157–175

Boudreau JC, Anderson W, Oravec J (1975) Chemical stimulus determinants of cat geniculate ganglion chemoresponsive group II unit discharge. Chem Senses Flavor 1 (4):495–518

Boudreau JC, Oravec JJ, Hoang NK (1982) Taste systems of goat geniculate ganglion. J Neurophysiol 48 (5):1226–1242

Braun JJ, Lasiter PS, Kiefer SW (1982) The gustatory neocortex of the rat. Physiol Psychol 10 (1):13–45

Brouwer JN, Hellekant G, Kasahara H, van der Wel H, Zotterman Y (1973) Electrophysiological study of the gustatory effects of the sweet proteins monellin and thaumatin in monkey, guinea pig and rat. Acta Physiol Scand 89:550–557

Buresova O (1978) Neocortico-amygdalar interaction in the conditioned taste aversion in rats. Activitas Nervosa Superior (Praha) 20:224–230

Cagan RH (1971) Biochemical studies of taste sensation I. Binding of 14C labeled sugars to bovine taste papillae. Biochim Biophys Acta 252:199–206

Cagan RH, Boyle AG (1984) Biochemical studies of taste sensation. XI. Isolation, characterization and taste ligand binding activity of plasma membranes from catfish taste tissue. Biochim Biophys Acta 799:230–237

Cagan RH, Morris RW (1979) Biochemical studies of taste sensations: binding to taste tissue of 3H labelled monellin, a sweet-tasting protein. Proc Nat Acad Sci 76 (4): 1692–1696
Caprio J (1978) Olfaction and taste in the channel catfish: an electrophysiological study of the responses to amino acids and derivatives. J Comp Physiol 123:357–371
Cardello AV (1978) Chemical stimulation of single human fungiform taste papillae: sensitivity profiles and locus of stimulation. Sensory Processes 2:173–190
Cardello AV (1981) Comparison of taste qualities elicited by tactile, electrical and chemical stimulation of single human taste papillae. Percept Psychophys 29 (2):163–169
Chang F-C T, Scott TR (1984) Conditioned taste aversions modify neural responses in the rat nucleus tractus solitarius. J Neurosci 4 (7):1850–1862
Chevreul ME (1824) Des différentes manières dont les corps agissent sur l'organe du goût. J Physiol Exp Pathol 4:127–131
Cohn C (1914) Organic flavours. The relation of chemical constitution to taste. Pharmazeutische Zentralhalle 55:735–747
Collings VB (1974) Human taste response as a function of locus of stimulation on the tongue and soft palate. Percept Psychophys 16 (1):69–174
Curtis DW, Stevens DA, Lawless HT (1984) Perceived intensity of the taste of sugar mixtures and acid mixtures. Chemical Senses 9 (2):107–120
Dastoli FR, Price S (1966) Sweet sensitive protein from bovine taste buds: isolation and assay. Science 154:905–907
Dickman JD, Sweazey RD, Smith DV (1982) Converging taste receptor inputs to the hamster solitary nucleus. Proceedings of the 4th annual meeting of the association for chemoreception sciences. Sarasota, Abstract, p 7
DiLorenzo PM, Schwartzbaum JS (1982a) Coding of gustatory information in the pontine parabrachial nuclei of the rabbit: magnitude of neural response. Brain Res 251:229–244
DiLorenzo PM, Schwartzbaum JS (1982b) Coding of gustatory information in the pontine parabrachial nuclei of the rabbit: temporal patterns of neural response. Brain Res 251:245–257
Doetsch GS, Erickson RP (1970) Synaptic processing of taste-quality information in the nucleus tractus solitarius of the rat. J Neurophysiol 33 (4):490–507
Dugas du Villard X, Van der Wel H, Brouwer JN (1980) Enhancement of the perceived sucrose sweetness in the rat by thaumatin. Chemical Senses 5 (2):93–98
Dugas du Villard X, Her C, Mac Leod P (1981) Qualitative discrimination of sweet stimuli: behavioural study on rats. Chem Senses 6 (2):143–148
Erickson RP (1963) Sensory neural patterns and gustation. In: Zotterman Y (ed) Olfaction and taste I. Macmillan, New York, pp 205–213
Erickson RP (1968) Stimulus coding in topographic and nontopographic afferent modalities: on the significance of the activity of individual sensory neurons. Physiol Rev 75 (6):447–465
Erickson RP (1982) The across-fibre pattern theory: an organizing principle for molar neural function. Contrib Sensory Physiol 6:79–110
Erickson RP (1983a) Taste: a time for re-evaluation. ECRO News Lett no 27:281–282
Erickson RP (1983b) Thomas Young: on a zen riddle and the single neuron basis of behavior. Lecture given at the Eastern physiological association meetings. April 8, Philadelphia
Erickson RP (1984) On the neural bases of behavior. Am Sci 72:233–241
Erickson RP, Covey E (1980) On the singularity of taste sensation. Physiol Behav 25:527–533
Erickson RP, Poulos DA (1973) On the qualitative aspect of the temperature sense. Brain Res 61:107–112
Erickson RP, Doetsch GS, Marshall DA (1965) The gustatory neural response function. J Gen Physiol 2:247–263

Erickson RP, Covey E, Doetsch GS (1980) Neuron and stimulus typologies in the rat gustatory system. Brain Res 196:513–519

Faull JR, Halpern BP (1972) Taste stimuli: time course of peripheral nerve response and theoretical models. Science 178:73–75

Faurion A (1982) Etude des mecanismes de la chimioréception du goût sucré. Thesis, Université Pierre et Marie Curie Paris VI

Faurion A (1983) Taste: a time for re-evaluation – reply. ECRO News Lett no 28:291–293

Faurion A, Bertrand B (1985) Differential inhibition of the sweet taste of 20 sweeteners in 5 human subjects by pronase E. (in preparation)

Faurion A, Mac Leod P (1982) Sweet taste receptor mechanisms. In: Birch GG, Parker KJ (eds) Nutritive sweeteners. Applied Science, Barking, England, pp 247–273

Faurion A, Saito S, Mac Leod P (1977) Experimental evidence for several receptor sites involved in sweet taste chemoreception. In: Le Magnen J, Mac Leod P (eds) Olfaction and taste VI. I R L, London, p 60 (Abstract)

Faurion A, Bonaventure L, Bertrand B, Mac Leod P (1979) Multiple approach of the sweet taste sensory continuum: psychophysical and electrophysiological data. In: Van der Starre H (ed) Olfaction and taste VII. I R L, London, p 86 (Abstract)

Faurion A, Saito S, Mac Leod P (1980) Sweet taste involves several distinct receptor mechanisms. Chemical Senses 5 (2):107–121

Ferrell F (1984) Gustatory nerve response to sugars in neonatal puppies. Neurosci Behav Rev 8:185–190

Fisher RS, Almli CR (1984) Postnatal development of sensory influences on lateral hypothalamic neurons of the rat. Dev Brain Res 12:55–75

Fishman IY (1957) Single fiber gustatory impulses in rat and hamster. J Cell Comp Physiol 49:319–334

Fonberg E (1969) Effects of small dorsomedial amygdala lesions on food intake and acquisition of instrumental alimentary reactions in dogs. Physiol Behav 4:739–743

Fox AL (1932) The relationship between chemical constitution and taste. Proc Nat Acad Sci 18:115–120

Frank M (1973) An analysis of hamster afferent taste nerve response functions. J Gen Physiol 61:588–618

Frank M, Pfaffmann C (1969) Taste nerve fibers: a random distribution of sensitivities to four tastes. Science 164:1183–1185

Frank ME, Hettinger TP (1984) Temporal properties of impulses in sucrose responsive hamster taste neurons. Proceedings of the 6th annual meeting of the association for chemoreception sciences. Sarasota abstract

Frijters JER, Oude Ophuis PAM (1983) The construction and prediction of psychophysical power functions for the sweetness of equiratio sugar mixtures. Perception 12:753–767

Frijters JER, DeGraaf C, Koolen HCM (1984) The validity of the equiratio taste mixture model investigated with sorbitol-sucrose mixtures. Chem Senses 9 (3):241–248

Funakoshi M, Ninomiya Y (1977) Neural code for taste quality in the thalamus of the dog. In: Katsuki Y, Sato M, Takagi S, Oomura Y (eds) Food intake and chemical senses. Japan Societies Press, Tokyo, pp 223–232

Funakoshi M, Ninomiya Y (1983) Relations between the spontaneous firing rate and taste responsiveness of the dog cortical neurons. Brain Res 262:155–159

Funakoshi M, Kasahara Y, Yamamoto T, Kawamura Y (1972) Taste coding and central perception. In: Schneider D (ed) Olfaction and taste IV. Wissenschaftliche Verlagsgesellschaft, Stuttgart, pp 336–342

Ganchrow JR, Erickson RP (1970) Neural correlates of gustatory intensity and quality. J Neurophysiol 33:768–783

Gent JF (1979) An exponential model for adaptation in taste. Sensory Processes 3:303–316

Gent JF, Bartoshuk LM (1983) Sweetness of sucrose, neohesperidin dihydrochalcone, and saccharin is related to genetic ability to taste the bitter substance 6-n-propylthiouracil. Chem Senses 7 (3/4):265–272

Giachetti I, Mac Leod P (1975) Cortical neuron responses to odours in the rat. In: Denton D, Coghlan JP (eds) Olfaction and taste V. Academic, New York pp 303–307

Giachetti I, Mac Leod P (1977) Olfactory input to the thalamus: evidence for a ventroposteromedial projection. Brain Res 125:166–169

Gillan DJ (1984) Evidence for peripheral and central processes in taste adaptation. Percept Psychophys 35 (1):1–4

Giza BK, Scott TR (1983) Blood glucose selectively affects taste-evoked activity in rat nucleus tractus solitarius. Physiol Behav 31:643–650

Glaser D (1979) Gustatory preference behaviour in primates. In: Kroeze JHA (ed) Proceedings of the 3rd symposium organised by ECRO: Preference and chemoreception, pp 51–61

Glaser D, Hellekant G, Brouwer JN, Van der Wel H (1978) The taste responses in primates to the proteins thaumatin and monellin and their phylogenetic implications. Folia Primatol 29:56–63

Glaser D, Hellekant G, Brouwer JN, Van der Wel H (1984) Effects of gymnemic acid on sweet taste perception in primates. Chem Senses 8 (4):367–374

Goldstein NI, Cagan RH (1982) Biochemical studies of taste sensation: monoclonal antibody against L-alanine binding activity of catfish taste epithelium. Proc Nat Acad Sci USA 79:7595–7597

Goto N, Yamamoto T, Kaneko M, Tomita H (1983) Primary pontine hemorrhage and gustatory disturbance: clinicoanatomic study. Stroke 14 (4):507–511

Grill HJ, Norgren R (1978) The taste reactivity test. II Mimetic responses to gustatory stimuli in chronic thalamic and chronic decerebrate rats. Brain Res 143:281–297

Haefeli RJ, Glaser D (1984) Gustatory threshold values of xylitol in primates. Folia Primatol 43:181–184

Halpern BP (1967) Chemotopic coding for sucrose and quinine hydrochloride in the nucleus of the fasciculus solitarius. In: Hayashi T (ed) Olfaction and taste II. John Hopkins, Baltimore, pp 549–562

Halpern BP, Nelson LM (1965) Bulbar gustatory response to anterior and posterior tongue stimulation in the rat. Am J Physiol 209:105–119

Halpern BP, Tapper DN (1971) Taste stimuli: quality coding time. Science 171:1256–1258

Hamilton RB, Norgren R (1984) Central projections of gustatory nerves in the rats. J Comp Neurol 222:560–577

Hänig DP (1901) Zur Psychophysik des Geschmacksinnes. Philosophische Studien (Leipzig) 17:576–623

Hara S (1955) Interrelationship among stimulus intensity, stimulated area and reaction time in the human gustatory sensation. Bull Tokyo Med Dent Univ 2:147–158

Harder DB, Whitney G, Frye P, Smith JC, Rashotte ME (1984) Strain differences among mice in taste psychophysics of sucrose octaacetate. Chemical Senses 9 (4):311–323

Harper HW, Jay JR, Erickson RP (1966) Chemically evoked sensations from single human taste papillae. Physiol Behav 1:319–325

Hayashi H (1976) Nerve impulse sequences correlated with the four primary taste qualities in rat. Tohoku J Exp Med 118:25–33

Hellekant G (1975) Different types of sweet receptors in mammals. In: Denton DA, Coghlan JP (eds) Olfaction and taste V. Academic, New York, pp 15–22

Hellekant G (1976) On the gustatory effects of gymnemic acid and miraculin in dog, pig and rabbit. Chem Senses Flavor 2:85–95

Hellekant G, Roberts TW (1983) Study of the effect of gymnemic acid on taste in hamster. Chem Senses 8 (2):195–202

Hellekant G, Hagstrom EC, Kasahara Y, Zotterman Y (1974) On the gustatory effects of miraculin and gymnemic acid in the monkey. Chem Senses Flavor 1:137–145

Hellekant G, Glaser D, Brouwer JN, Van der Wel H (1976) Gustatory effects of miraculin, monellin and thaumatin in the Saguinus midas tamarin monkey studied with electrophysiological and behavioural techniques. Acta Physiol Scand 97:241–250

Hellekant G, Glaser D, Brouwer J, Van der Wel H (1981) Gustatory responses in three prosimian and two simian primate species (Tupaia glis, Nycticebus coucang, Galago senegalensis, Callithrix j. Jacchus and Saguinus midas niger) to six sweeteners and miraculin and their phylogenetic implications. Chem Senses 6 (3):165–173

Henning H (1916) Die Qualitätenreihe des Geschmacks. Z Psychol Z Angew Psychol Charakterkd 74:203–219

Hermann GE, Kohlerman NJ, Rogers RC (1983) Hepatic-vagal and gustatory afferent interactions in the brainstem of the rat. J Auton Nerv System 9, 477–495

Herrick CJ (1905) The central gustatory paths in the brains of bony fishes. J Comp Neurol 15:375–456

Hidaka I (1970) The effects of transition metals on the palatal chemoreceptors of the carp. Jpn J Physiol 20:599–609

Hiji Y (1975) Selective elimination of taste response to sugars by proteolytic enzymes. Nature 256:427–429

Hiji Y, Ito J (1977) Removal of sweetness by proteases and its recovery mechanism in rat taste cells. Comp Biochem Physiol 58 (1A):109–113

Hiji Y, Sato M (1972) Properties of sweet sensitive protein extracted from the rat tongue. In: Schneider D (ed) Olfaction and taste IV. Wissenschaftliche Verlagsgesellschaft, Stuttgart, pp 221–225

Hiji Y, Sato M (1973) Isolation of the sugar-binding protein from rat taste buds. Nature 244:91–93

Hiji Y, Kobayashi N, Sato M (1969) Kumamoto Med J 22:104–107

Hiji Y, Kobayashi N, Sato M (1971) Sweet-sensitive protein from the rat tongue: its interaction with various sugars. Comp Biochem Physiol [B] 39:367–375

Hill AV (1913) The combinations of haemoglobin with oxygen and with carbon mono-oxide I. Biochem J 7:471–480

Hopkins CY (1942) Taste differences in compounds having the NCS linkage. Can J Res 20B:268–273

Hornung DE, Enns MP (1984) The independence and integration of olfaction and taste. Chem Senses 9 (2):97–106

Hough CAM, Edwardson JA (1978) Antibodies to thaumatin as a model of the sweet taste receptor. Nature 271:381–383

Hyman AM, Frank ME (1980a) Effect of binary taste stimuli on the neural activity of the hamster chorda tympani. J Gen Physiol 76:125–142

Hyman AM, Frank ME (1980b) Sensitivities of single nerve fibers in the hamster chorda tympani to mixtures of taste stimuli. J Gen Physiol 76:143–173

Ichioka M (1972) Neural correlates of taste sensation quality. Experientia 28:523–524

Ichioka M, Hayashi H (1974) Spatio-temporal nerve impulse patterns in rat chorda tympani fibres in correlation with four primary taste qualities. Proc Jpn Acad 50:392–395

Iggo A, Leek BF (1967) The afferent innervation of the tongue of the sheep. In: Hayashi T (ed) Olfaction and taste II. Pergamon, London, pp 493–507

Imoto T, Yamada H, Hiji Y (1981) Flourescent labeling of the proteins on taste papillae. In: Hiji Y (ed) Proceedings of the 15th Japanese symposium on taste and smell. Osaka University, Osaka, pp 206–209

Ishiko N (1974) Local gustatory functions associated with segmental organization of the anterior portion of the cat's tongue. Exp Neurol 45:341–354

Ishiko N, Akagi T (1972) Topographical organization of gustatory nervous system. In: Schneider D (ed) Olfaction and taste IV. Wissenschaftliche Verlagsgesellschaft, Stuttgart, pp 343–439
Iwasaki K, Sato M (1984a) Inhibitory effects of some heavy metal ions on taste nerve responses in mice. Jpn J Physiol 34:907–918
Iwasaki K, Sato M (1984b) Neural and behavioral responses to taste stimuli in the mouse. Physiol Behav 32:803–807
Jakinovich W (1981) Stimulation of the gerbil gustatory receptors by artificial sweeteners. Brain Res 210:69–81
Jakinovich W (1982) Taste aversion to sugars by the gerbil. Physiol Behav 28:1065–1071
Jakinowich W Jr (1983) Methyl 4, 6-dichloro-4, 6-dideoxy-α-D-galactopyranoside: an inhibitor of sweet taste responses in gerbils. Science 219:408–410
Kadowaki H, Sasaki H, Yamada H, Hiji Y (1979) Analysis of the unique protein obtained from fungiform or circumvallate papillae of man, monkey and dog. In: Funakoshi M (ed) Taste and smell XIII. GIFU, Japan, pp 92–94
Kalmus H (1971) Genetics of taste. In: Beidler LM (ed) Taste. Handbook of sensory physiology, vol 4. Springer, Berlin Heidelberg New York, pp 165–179
Kang R, Herman D, Macgillis M, Zarzecki P (1985) Convergence of sensory inputs in somatosensory cortex: interactions from separate afferent sources. Exp Brain Res 57:271–278
Kashiwagura T, Kamo N, Kurihara K, Kobatake Y (1976) Phasic and tonic components of gustatory responses in the frog. Am J Physiol 231:1097–1104
Kashiwagura T, Kamo N, Kurihara K, Kobatake Y (1977) Responses of the frog gustatory receptors to various odorants. Comp Biochem Physiol [C] 56:105–108
Kemble ED, Schwartzbaum JS (1969) Reactivity of taste properties of solutions following amygdaloid lesions. Physiol Behav 4:981–985
Kennedy LM, Halpern BP (1980a) Fly chemoreceptors: a model for the taste modifier ziziphin. Physiol Behav 24:135–143
Kennedy LM, Halpern BP (1980b) Extraction, purification and characterization of a sweetness-modifying component from *Ziziphus jujuba*. Chem Senses 5:1379–1384
Kennedy LM, Sturckow B, Waller FJ (1975) Effect of gymnemic acid on single taste hairs of the housefly Musca domestica. Physiol Behav 14:755–765
Kiefer SW, Leach LR, Braun JJ (1984) Taste agnosia following gustatory neocortex ablation: dissociation from odor and generality across taste qualities. Behav Neurosci 98 (4):590–608
Kier LB (1972) A molecular theory of sweet taste. J Pharm Sci 61 (9):1343–1397
Kiesow F (1898) Expériences gustatives sur diverses papilles isolément excitées. Arch Ital Biol 30:399–425
Kimura K, Beidler LM (1961) Microelectrode study of taste receptors of rat and hamster. J Cell Comp Physiol 58:131–139
Kitada Y (1984) Two different receptor sites for Ca2+ and Na+ in frog taste responses. Neurosci Lett 47:63–68
Köhler W (1930) Gestalt Psychology. Gallimard, Paris (1964)
Koyama N, Kurihara K (1971) Do unique proteins exist in taste buds? J Gen Physiol 57:297–302
Kroeze JHA (1979) Masking and adaptation of sugar sweeteness intensity. Physiol Behav 22:347–351
Kroeze JHA (1982a) After repetitious sucrose stimulation saltiness suppression in NaCl-sucrose mixtures is diminished: implications for a central mixture suppression mechanism. Chem Senses 7 (1):81–92
Kroeze JHA (1982b) The influence of relative frequence of pure and mixed stimuli on mixture suppression in taste. Percept Psychophys 31:276–278

Kroeze JHA (1983) Successive contrast cannot explain suppression release after repetitious exposure to one of the components of a taste mixture. Chem Senses 8 (2):211–223

Kubota T, Kubo I (1969) Bitterness and chemical structure. Nature 233:97–99

Kuznicki JT (1978) Taste profiles from simple human taste papillae. Percept Mot Skills 47:279–286

Landgren S (1957) Convergence of tactile, thermal, gustatory impulses on single cortical cells. Acta Physiol Scand 40:210–221

Lasiter PS (1982) Cortical substrates of taste aversion learning: direct amygdaloid projections to the gustatory neocortex do not mediate conditioned taste aversion learning. Physiol Psychol 10 (4):377–383

Lasiter PS, Glanzman DL (1983) Axon collaterals of pontine taste area neurons project to the posterior ventromedial thalamic nucleus and to the gustatory neocortex. Brain Res 258:299–304

Lasiter PS, Glanzman DL, Mensah PA (1982) Direct connectivity between pontine taste areas and gustatory neocortex in rat. Brain Res 234:111–121

Lawless H (1980) A comparison of different methods used to assess sensitivity to the taste of phenylthiocarbamide (PTC). Chem Senses 5 (3):247–256

Lawless HT, Stevens DA (1983) Cross adaptation of sucrose and intensive sweeteners. Chem Senses 7 (3–4):309–315

Lawless H, Stevens DA (1984) Effects of oral chemical irritation on taste. Physiol Behav 32:995–998

Lush IE (1981) The genetics of tasting in mice. I. Sucrose octaacetate. Genet Res 38:93–95

Lush IE (1982) The genetics of tasting in mice. II. Strychnine. Chem Senses 7 (1):93–98

Maes FW (1984) A neural coding model for sensory intensity discrimination, to be applied to gustation. J Comp Physiol [A] 155:263–270

Maes FW, Erickson RP (1984) Gustatory intensity discrimination in rat NTS: a tool for the evaluation of neural coding theories. J Comp Physiol [A] 155:271–282

Maga JA (1983) Flavor potentiators. CRC Crit Rev Food Sci Nutr 18 (3):231–312

Makous W, Nord S, Oakley B, Pfaffmann C (1963) The gustatory relay in the medulla. In: Zotterman Y (ed) Olfaction and taste I. Macmillan, New York

Marks LE (1978) The unity of the senses. Interrelations among the modalities. Academic, New York

Marui T, Harada S, Kasahara Y (1983a) Gustatory specificity for amino acids in the facial taste system of the carp, Cyprinus carpio L. J Comp Physiol 153:299–308

Marui T, Evans RE, Zielinski B, Hara TJ (1983b) Gustatory responses of the rainbow trout (Salmo gairdneri) palate to amino acids and derivatives. J Comp Physiol 153:423–433

Matsuo R, Kusano K (1984) Lateral hypothalamic modulation of the gustatory-salivary reflex in rats. J Neurosci 4 (5):1208–1216

Matsuo R, Shimizu N, Kusano K (1984) Lateral hypothalamic modulation of oral sensory afferent activity in nucleus tractus solitarius neurons of rats. J Neurosci 4 (5):1201–1207

McBride RL (1984) The sweetness of binary mixtures of sucrose, fructose, and glucose. In: Proceedings of the 6th annual meeting of the association for chemoreception sciences. Sarasota (Abstract)

McBurney DH (1972) Gustatory cross adaptation between sweet-tasting compounds. Percept Psychophys 11:225–227

McBurney DH, Gent JF (1978) Taste of methyl alpha-D-mannopyranoside: effects of cross adaptation and Gymnema sylvestre. Chem Senses Flavour 3 (1):945–950

McBurney DH, Bartoshuk LM (1973) Interactions between stimuli with different taste qualities. Physiol Behav 10:101–106

McBurney DH, Smith DV, Shick TR (1972) Gustatory cross adaptation: sourness and bitterness. Percept Psychophys 11 (3):228–232

McBurney DH, Collings VB, Glanz LM (1973) Temperature dependence of human taste responses. Physiol Behav 11:89–94

McCutcheon NB, Saunders J (1972) Human taste papilla stimulation: stability of quality judgements over time. Science 175:214–216
Mehren MJ, Church DC (1976) Influence of taste-modifiers on taste responses to amino-acids. Anim Product 22:255–260
Meiselman HL, Halpern BP (1970) Effects of Gymnema sylvestre on complex tastes elicited by amino acids and sucrose. Physiol Behav 5:1379–1384
Meiselman HL, Bose HE, Nykvist WE (1972) Effect of flow rate on tase intensity responses in humans. Physiol Behav 9:35–38
Meiselman HL, Halpern BP, Dateo GP (1976) Reduction of sweetness judgments by extracts from the leaves of Ziziphus jujuba. Physiol Behav 17:313–337
Miller IJ Jr (1971) Peripheral interactions among single papilla inputs to gustatory nerve fibres. J Gen Physiol 57:1–25
Miller IJ Jr (1974) Branched chorda tympani neurons and interactions among taste receptors. J Comp Neurol 158 (2):155–166
Miller IJ Jr (1977) Gustatory receptors of the palate. In: Sato M, Takagi S, Oomura Y (eds) Food intake and chemical senses. Japan Scientific Societies Press, Tokyo, pp 173–185
Miller IJ Jr, Spangler KM (1982) Taste bud distribution and innervation on the palate of the rat. Chem Senses 7 (1):99–108
Mistretta CM (1972): A quantitative analysis of rat chorda tympani fiber discharge patterns. In: Schneider D (ed) Olfaction and taste IV. Wissenschaftliche Verlagsgesellschaft, Stuttgart, pp 294–300
Miyake M, Kamo N, Kurihara K, Kobatake Y (1976) Physicochemical studies of taste reception: suppressive effect of salts on sugar response of the frog. Biochim Biophys Acta 436:856–862
Monod J, Wyman J, Changeux JP (1965) On the nature of allosteric transitions: a plausible model. J Mol Biol 12:88–118
Mooser G (1980) Sodium and potassium salt stimulation of taste receptor cells: an allosteric model. Proc Nat Acad Sci, USA 77 (3):1686–1690
Morrison GR, Jessup A (1977) A dual taste of saccharin in the rat. Chem Senses Flavour 2:395–400
Moskowitz HR (1973) Effects of solution temperature on taste intensity in humans. Physiol Behav 10:289–292
Mozell MM, Smith BP, Smith PE, Sullivan RL, Swender P (1969) Nasal chemoreception in flavor identification. Arch Otolaryngol 90:131–137
Murphy C, Cain WS (1980) Taste and olfaction: independence vs interaction. Physiol Behav 24:601–605
Murphy C, Cardello AV, Brand JG (1981) Tastes of fifteen halide salts following water and NaCl: anion and cation effects. Physiol Behav 26:1083–1095
Nachman M, Ashe JA (1974) Effects of basolateral amygdala lesions on neophobia, learned taste aversions and sodium appetite in rats. J Comp Physiol Psychol 87:622–643
Nagai T, Ueda K (1981) Stochastic properties of gustatory impulse discharges in rat chorda tympani fibers. J Neurophysiol 45 (3):574–592
Nagy JI, Goedert M, Hunt SP, Bond A (1982) The nature of the substance P-containing nerve fibres in taste papillae of the rat tongue. Neuroscience 7:3137–3151
Narahashi Y (1970) Pronase. In: Perlmann GE, Lorrand L (eds) Methods in enzymology, XIX. Academic, New York, pp 651–664
Nilsson B (1979) Taste acuity of the human palate III. Studies with taste solutions on subjects in different age groups. Acta Odontol Scand 37:235–252
Ninomiya Y, Funakoshi M (1982a) Relationships between spontaneous discharge rates and taste responses of the dog thalamic neurons. Brain Res 242:67–76
Ninomiya Y, Funakoshi M (1982b) Responsiveness of dog thalamic neurons to taste stimulation of various tongue regions. Physiol Behav 29:741–745

Ninomiya Y, Mizukoshi T, Higashi T, Katsukawa H, Funakoshi M (1984) Gustatory neural responses in three different strains of mice. Brain Res 302:305–314

Noma A, Hiji Y (1972) Effects of modifiers on taste responses in the rat chorda tympani. Jpn J Physiol 22:393–402

Noma A, Sato M, Tsuzuki Y (1974) Taste effectiveness of anomers of sugars and glycosides as revealed from hamster taste responses. Comp Biochem Physiol A 48:249–262

Nomoto M, Naharashi Y, Murakami M (1960) A proteolytic enzyme of Streptomyces griseus. VI. Hydrolysis of protein by Streptomyces griseus protease. J Biochem (Tokyo) 48 (4):593–602

Nomura S, Mizuno N, Itoh K, Matsuda K, Sugimoto T, Nakamura Y (1979) Localization of parabrachial nucleus neurons projecting to the thalamus or the amygdala in the cat using horseradish peroxydase. Exp Neurol 64:375–385

Norgren R (1970) Gustatory responses in the hypothalamus. Brain Res 21:63–77

Norgren R (1971) Taste pathways in rat brain stem. Science 173:1136–1139

Norgren R (1976) Taste pathways to hypothalamus and amygdala. J Comp Neurol 166:17–30

Norgren R (1977) A synopsis of gustatory neuroanatomy. In: Le Magnen J, Mac Leod P (eds) Olfaction and taste VI. I R L London, pp 225–232

Norgren R (1978) Projections from the nucleus of the solitary tract in the rat. Neuroscience 3:207–218

Norgren R (1981) The central organization of the gustatory and visceral afferent systems in the nucleus of the solitary tract. In: Katsuki Y, Norgren R, Sato M (eds) Brain mechanisms of sensation. John Wiley, New York, pp 143–160

Norgren R, Leonard CM (1971a) Taste pathways in rat brainstem. Science 173:1136–1139

Norgren R, Leonard CM (1971b) Ascending central gustatory pathways. J Comp Neurol 150:217–238

Norgren R, Wolf G (1975) Projections of thalamic gustatory and lingual areas in the rat. Brain Res 92:123–129

Nowlis GH, Frank ME, Pfaffmann C (1980) Specificity of acquired aversions to taste qualities in hamsters and rats. J Comp Physiol Psychol 94:932–942

O'Mahony M, Buteau L (1982) Taste mixtures: can the components be readily identified? IRCS Med Sci 10:9–10

O'Mahony M, Atassi-Sheldon S, Rothman L, Murphy-Ellison T (1983) Relative singularity/mixedness judgements for selected taste stimuli. Physiol Behav 31:749–755

Oakley B (1967) Altered temperature and taste responses from cross regenerated sensory nerves in the rat's tongue. J Physiol 188:353–371

Oakley B (1970) Reformation of taste buds by crossed sensory nerves in the rat's tongue. Acta Physiol Scand 79:88–94

Oakley B (1974) On the specification of taste neurons in the rat tongue. Brain Res 75:85–96

Oakley B (1975) Receptive fields of cat taste fibers. Chem Senses Flavour 1:431–442

Oakley B, Chu JS, Jones LB (1981) Axonal transport maintains taste responses. Brain Res 221:289–298

Oakley B, Keppel E, Hugues SE (1984) Trophic capacity of experimentally lengthened gustatory axons. Dev Brain Res 16:195–201

Ogawa H, Hayama T (1984) Receptive fields of solitario-parabrachial relay neurons responsive to natural stimulation of the oral cavity in rats. Exp Brain Res 54:359–366

Ogawa H, Kaisaku J (1980) Projection to parabrachial nucleus of solitary tract nucleus neurons activated by tongue afferents in rats. Jpn J Physiol 30:659–663

Ogawa H, Sato M, Yamashita S (1968) Multiple sensitivity of chorda tympani fibers in the rat and hamster to gustatory and thermal stimuli. J Physiol 199:223–240

Ogawa H, Sato M, Yamashita S (1969) Gustatory impulse discharges in response to saccharin in rats and hamsters. J Physiol (Lond) 204:311–329

Ogawa H, Sato M, Yamashita S (1973) Variability in impulse discharges in rat chorda tympani fibers in response to repeated gustatory stimulations. Physiol Behav 11:469–479
Ogawa H, Yamashita S, Sato M (1974) Variation in gustatory nerve fiber discharge pattern with change in stimulus concentration and quality. J Neurophysiol 37 (3):443–457
Ogawa H, Imoto T, Hayama T (1980) Taste relay neurons in the solitary tract nucleus of the rats. Neurosci Lett 18:295–299
Ogawa H, Hayama T, Ito S (1981) Receptive fields of neurons in the brain stem taste nuclei of rats. In: Hiji Y (ed) Proceedings of the 15th Japanese symposium on taste and smell. Tottori University, Japan, pp 137–140
Ogawa H, Hayama T, Ito S (1982) Convergence of input from tongue and palate to the parabrachial nucleus neurons of rats. Neurosci Lett 28:9–14
Ogawa H, Imoto T, Hayama T (1984) Responsiveness of solitario-parabrachial relay neurons to taste and mechanical stimulation applied to the oral cavity in rats. Exp Brain Res 54:349–358
Ostretsova IB, Safarian EK, Etingof RN (1975) On the presence and localisation of proteins binding glucose in the tongue. Proc Acad Sci USSR 233:1484–1487
Ozeki M, Sato M (1972) Responses of gustatory cells in the tongue of rat to stimuli representing the four taste qualities. Comp Biochem Physiol [A] 41:391–407
Pangborn RM, Chrisp RB, Bertolero LM (1970) Gustatory, salivary, and oral thermal responses to solution of sodium chloride at four temperatures. Percept Psychophys 8:69–75
Paulus K, Reisch AM (1980) The influence of temperature on the threshold values of primary tastes. Chem Senses 5 (1):11–21
Perrotto RS, Scott TS (1976) Gustatory neural coding in the pons. Brain Res 110:283–300
Pfaffmann C (1939) Specific gustatory impulses. J Physiol 96:41P–42P
Pfaffmann C (1941) Gustatory afferent impulses. J Cell Comp Physiol 17:243–258
Pfaffmann C (1955) Gustatory nerve impulses in rat, cat and rabbit. J Neurophysiol 18:429–440
Pfaffmann C (1959) The sense of taste. In: Field IJ, Magoun HW, Hall VE (eds) (Handbook of physiology, sect 1) Neurophysiology, vol 1. American Physiological Society, Washington, pp 507–534
Pfaffmann, C (1974) Specificity of the sweet receptors of the squirrel monkey. Chem Senses Flavour 1 (1):61–67
Pfaffmann C, Norgren R, Grill HJ (1977) Sensory affect and motivation. Ann N Y Acad Sci 290:18–34
Pfaffmann C, Frank M, Norgren R (1979) Neural mechanisms and behavioral aspects of taste. Ann Rev Psychol 30:283–325
Pritchard TC, Scott TR (1982) Amino acids as taste stimuli. II Quality coding. Brain Res 253:93–104
Rifkin B, Bartoshuk LM (1980) Taste synergism between monosodium glutamate and disodium 5′ guanylate. Physiol Behav 24:1169–1172
Riskey DR, Desor JA, Vellucci D (1982) Effects of gymnemic acid concentration and time since exposure on intensity of simple tastes: a test of the biphasic model for the action of gymnemic acid. Chem Senses 7 (2):143–151
Rolls ET, Rolls BJ (1977) Activity of neurones in sensory, hypothalamic and motor areas during feeding in the monkey. In: Katsuki Y, Sato M, Takagi S, Oomura Y (eds) Food intake and chemical senses. Jpn Sci Soc Tokyo, pp 525–549
Rolls ET, Burton MJ, Mora F (1976) Hypothalamic neuronal responses associated with the sight of food. Brain Res 111:53–66
Sandick B, Cardello AV (1981) Taste profiles from single circumvallate papillae: comparison with fungiform profiles. Chem Senses 6 (3):197–214
Sato M (1963) The effect of temperature change on the response of taste receptors. In: Zotterman Y (ed) Olfaction and taste I. Pergamon Oxford, pp 151–164

Sato M, Ogawa H, Yamashita S (1975) Response properties of macaque monkey chorda tympani fibers. J Gen Physiol 66:781–810
Sato M, Hiji Y, Ito H (1977a) Taste discrimination in the monkey. In: Le Magnen J. Mac Leod P (eds) Olfaction and taste VI. I R L, London, pp 233–240
Sato M, Hiji Y, Ito H, Imoto T (1977b) In: Kare MR, Maller O (eds) Chemical senses and nutrition. Academic, New York, pp 327–341
Sato M, Hiji Y, Ito H, Imoto T, Saku C (1977c) Properties of sweet taste receptors in macaque monkeys. In: Katsuki Y, Sato M, Takagi SF, Oomura Y (eds) Food intake and chemical senses. Jpn Sci Soc, Tokyo, pp 187–199
Sato T (1971) Site of gustatory neural adaptation. Brain Res 34:385–388
Sato T (1977) An initial phasic depolarization exists in the receptor potential of taste cells. Experientia 33:1165–1167
Sato T (1980) Recent advances in the physiology of cells. Prog Neurobiol 14:25–67
Sato T, Beidler LM (1975) Membrane resistance change of the frog taste cells in response to water and NaCl. J Gen Physiol 66:735–763
Schiffman SS, Erickson RP (1971) A psychophysic model for gustatory quality. Physiol Behav 7:617–633
Schiffman SS, Erickson RP (1980) The issue of primary tastes versus a taste continuum. Neurosci Behav Rev 4:109–117
Schiffman SS, Reilly DA, Clark TB (1979) Qualitative differences among sweeteners. Physiol Behav 23:1–9
Schiffman SS, McElroy AE, Erickson RP (1980) The range of taste quality of sodium salts. Physiol Behav 24:217–224
Schiffman SS, Cahn H, Lindley MG (1981) Multiple receptor sites mediate sweetness: evidence from cross adaptation. Pharmacol Biochem Behav 15:377–388
Schwartzbaum JS, Block CH (1981) Interrelations between parabrachial pons and ventral forebrain of rabbits in taste-mediated functions. In: Ben-Ari Y (ed) The amygdaloid complex, INSERM symposium no 20. Elsevier, North-Holland Biomedical, Amsterdam, pp 367–382
Schwartzbaum JS, DiLorenzo PM (1982) Gustatory functions of the nucleus tractus solitarius in the rabbit. Brain Res Bull 8:285–292
Scott TR, Chang F-CT (1984) The state of gustatory neural coding. Chem Senses 8 (3):297–314
Scott TR, Erickson RP (1971) Synaptic processing of taste quality information in the thalamus of the rat. J Neurophysiol 34 (5):868–884
Scott TR, Perrotto RS (1980) Intensity coding in pontine taste area: gustatory information is processed similarly throughout rat's brainstem. J Neurophysiol 44 (4):739–750
Scott TR, Pritchard TC (1982) A neurobehavioral analysis of l-aminoacids as taste stimuli. Perfumer Flavorist 7 (1):1–25
Scott TR, Yalowitz MS (1978) Thalamic taste responses to changing stimulus concentration. Chem Senses Flavour 3 (2):167–175
Shallenberger RS, Acree TE (1967) Molecular theory of sweet taste. Nature 216:480–482
Shallenberger RS, Acree TE (1971) Chemical structure of compounds and their sweet and bitter taste. In: Beidler LM (ed) Handbook of sensory physiology, vol 4. Springer, Berlin Heidelberg New York, pp 221–273
Shimazaki K, Sato M (1981) Binding of sweeteners to a protein fraction from monkey taste papillae. In: Hiji Y (ed) Proceedings of the 15th symposium on taste and smell. Osaka University, Japan, pp 202–205
Shimazaki K, Sato M, Takegami T (1981) Binding of 35S saccharin to a protein fraction of rat tongue epithelia. Biochim Biophys Acta 677:331–338
Shipley MT, Geinisman Y (1984) Anatomical evidence for convergence of olfactory, gustatory, and visceral afferent pathways in mouse cerebral cortex. Brain Res Bull 12:221–226

Sloan HE, Hugues SE, Oakley B (1983) Chronic impairment of axonal transport eliminates taste responses and taste buds. J Neurosci 3 (1):117–123

Smith DV, Bealer SL (1976) Recovery of excitability after gustatory adaptation: effects of stimulus intensity. Sensory Processes 1:99–108

Smith DV, Frank M (1972) Cross adaptation between salts in the chorda tympani nerve of the rat. Physiol Behav 8:213–220

Smith DV, Steadman JW, Rhodine CN (1975) An analysis of the time course of gustatory neural adaptation in the rat. Am J Physiol 229 (4):1134–1140

Smith DV, Bealer SL, Van Buskirk RL (1978) Adaptation and recovery of the rat chorda tympani response to NaCl. Physiol Behav 20:629–636

Smith DV, Travers JB, Van Buskirk RL (1979) Brainstem correlates of gustatory similarity in the hamster. Brain Res Bull 4:359–372

Smith DV, van Buskirk RL, Travers JB, Bieber SL (1983a) Gustatory neuron types in hamster brain stem. J Neurophysiol 50 (2):522–540

Smith DV, van Buskirk RL, Travers JB, Bieber SL (1983b) Coding of taste stimuli by hamster brain stem neurons. J Neurophysiol 50 (2):541–558

Smith VV, Halpern BP (1983) Selective suppression of judged sweetness by ziziphins. Physiol Behav 30:867–874

Steiner JE (1973) The gustofacial response: observation on normal and anencephalic newborn infants. In: Bosma JF (ed) IVth Symposium on oral sensation and perception. DHEW, Bethesda, pp 254–278

Tateda H (1967) Sugar receptor and alpha-aminoacids in the rat. In: Hayashi T (ed) Olfaction and taste II. Pergamon, Oxford, pp 383–397

Tonosaki K, Funakoshi M (1984a) Intracellular taste cell responses of mouse. Comp Biochem Physiol 78A (4):651–656

Tonosaki K, Funakoshi M (1984b) The mouse taste cell response to five sugar stimuli. Comp Biochem Physiol 79A (4):625–630

Travers JB, Smith DV (1979) Gustatory sensitivities in neurons of the hamster nucleus tractus solitarius. Sensory Processes 3:1–26

Travers SP, Smith DV (1984) Responsiveness of neurons in the hamster parabrachial nuclei to taste mixtures. J Gen Physiol 84:221–250

Van Buskirk RL, Smith DV (1981) Taste sensitivity of hamster parabrachial pontine neurons. J Neurophysiol 45 (1):144–171

Van Buskirk RL, Erickson RP (1977) Odorant responses in taste neurons of the rat NTS. Brain Res 135:287–303

Von Bekesy G (1965) The effect of adaptation on the taste threshold observed with a semiautomatic gustometer. J Gen Physiol 48:81–88

Von Bekesy G (1966) Taste theories and the chemical stimulation of single papillae. J Appl Physiol 21:1–9

Warren RM, Pfaffmann C (1959) Suppression of sweet sensitivity by potassium gymnemate. J Appl Physiol 14 (1):40–42

West CHK, Bernard RA (1978) Intracellular characteristics and responses of taste buds and lingual cells of the mudpuppy. J Gen Physiol 72:305–326

Whitehead MC, Frank ME (1983) Anatomy of the gustatory system in the hamster: central projections of the chorda tympani and the lingual nerve. J Comp Physiol 220:378–395

Wieser H, Belitz HD (1975) Zusammenhänge zwischen Struktur und Bittergeschmack bei Aminosäuren und Peptiden I. Aminosäuren und verwandte Verbindungen. Z Lebensm Unters Forsch 159:65–72

Woolston DC, Erickson RP (1979) Concept of neuron types in gustation in the rat. J Neurophysiol 42 (5):1390–1409

Yackzan KS (1966) Biological effects of Gymnema sylvestre fractions. Ala J Med Sci 3:1–9

Yackzan KS (1969) Biological effects of Gymnema sylvestre fractions II. Electrophysiology. Effect of gymnemic acid on taste receptor response. Ala J Med Sci 6:455–463

Yamaguchi S (1967) The synergistic taste effect of monosodium glutamate and disodium 5′-inosinate. J Food Sci 32 (14):473–478

Yamaguchi S (1979) The umami taste. In: Boudreau JC (ed) Food and taste chemistry. University of Texas, Houston, pp 33–52

Yamaguchi S, Takahashi C (1984) Interactions of monosodium glutamate and sodium chloride on saltiness and palatability of a clear soup. J Food Sci 49 (1):82–85

Yamamoto T (1983) Neural mechanisms of taste function. Front Oral Physiol 4:102–130

Yamamoto T, Kawamura Y (1971) Inhibitory effect of cupric and zinc ions on sweet taste response in the rat. J Osaka University Dental School 11:99–104

Yamamoto T, Kawamura Y (1972) A model of neural code for taste quality. Physiology Behav 9:559–563

Yamamoto T, Kawamura Y (1975) Dual innervation of the foliate papillae of the rat: an electrophysiological study. Chem Senses Flavor 1:241–244

Yamamoto T, Kawamura Y (1974) An off-type response of the chorda tympani nerve in the rat. Physiol Behav 13:239–243

Yamamoto T, Kawamura Y (1977) Physiological characteristics of cortical taste area. In: Le Magnen J, Mac Leod P (eds) Olfaction and taste VI. IRL, London, pp 257–264

Yamamoto T, Kawamura Y (1984) Gustatory reaction time to various salt solutions in human adults. Physiol Behav 32:49–53

Yamamoto T, Matsuo R, Kawamura Y (1980a) Localization of cortical gustatory area in rats and its role in taste discrimination. J Neurophysiol 44 (3):40–45

Yamamoto T, Matsuo R, Kawamura Y (1980b) The pontine taste area in the rabbit. Neurosci Lett 16:5–9

Yamamoto T, Matsuo R, Kawamura Y (1980c) Corticofugal effects on the activity of thalamic taste cells. Brain Res 193:258–262

Yamamoto T, Azuma S, Kawamura Y (1981a) Significance of cortical-amygdalar-hypothalamic connections in retention of conditioned taste aversion in rats. Exp Neurol 74:758–768

Yamamoto T, Takahashi T, Kawamura Y (1981b) Access to the cerebral cortex of extralingual taste inputs in the rat. Neurosci Lett 24:129–132

Yamamoto T, Yuyama N, Kawamura Y (1981c) Cortical representation of sensory modalities from the tongue in the rat. In: Kawamura Y, Dubner R (eds) Oral-facial sensory and motor functions. Quintessence, Tokyo

Yamamoto T, Yuyama N, Kawamura Y (1981d) Cortical neurons responding to tactile, thermal and taste stimulation of the rat's tongue. Brain Res 221:202–206

Yamamoto T, Yuyama N, Kawamura Y (1981e) Central processing of taste perception. In: Katsuki Y, Sato M, Norgren R (eds) Brain mechanisms of sensation. John Wiley, New York

Yamamoto T, Kato T, Kawamura Y, Yoshida M (1983) Gustatory reaction time to various sweeteners. In: Kasahara Y et al. (eds) Proceedings of the 17th Japanese symposium on taste and smell. Kagoshima, 1983. Osaka University, p 80

Yamamoto T, Azuma S, Kawamura Y (1984a) Functional relations between the cortical gustatory area and the amygdala: electrophysiological and behavioral studies in the rat. Exp Brain Res 56:23–31

Yamamoto T, Matsuo R, Kawamura Y (1984b) Reaction time to taste stimuli applied to one side of the human tongue. In: Ueda K (ed) Proceedings of the 18th Japanese symposium on taste and smell. Tokyo. Osaka University, p 172

Yamamoto T, Yuyama N, Kato T, Kawamura Y (1984c) Gustatory responses of cortical neurons in rats. I. Response characteristics. J Neurophysiol 51 (4):616–635

Yamashita S, Sato M (1965) The effect of temperature on gustatory response of rats. J Cell Comp Physiol 66:1–18

Yamashita S, Ogawa H, Kiyohara T, Sato M (1970) Modification by temperature change of gustatory impulse discharges in chorda tympani fibres of rats. Jpn J Physiol 20:348–363

Yoshida M, Saito S (1969) Multidimensional scaling of the taste of aminoacids. Jpn Psychol Res 11 (4):149–166

Yoshii K, Kobatake Y, Kurihara K (1981) Selective enhancement and suppression of frog gustatory responses to aminoacids. J Gen Physiol 77:373–385

Zawalich WS (1973) Depression of gustatory sweet response by alloxan. Comp Biochem Physiol [A] 44: 903–909

Zotterman Y (1935) Action potentials in the glossopharyngeal nerve and in the chorda tympani. Skand Arch Physiol 72:73–77

Addendum

Since this paper was written, others have been published which update some points developed therein:

Concerning the laterality (cf. Sect. 2.1.7, p. 138) of taste projections, Ogawa et al. (Exp Neurol 83:507–517, 1984) found that the taste-responding neurons of PBN were possibly activated by ipsi-, contra-, or bilateral antidromic stimulation of the thalamic taste area. Nomura and Ogawa (Neurosci Res 3:91–105, 1985) did not confirm the ipsilateral projection to the thalamus VPMpc area. They found bilateral projections of the "oral cavity" with widespread and complex receptive fields. They nonetheless confirmed the plurimodal responsiveness of thalamic taste neurons. Ogawa et al. (Neurosci Res 2:447–459, 1985) also confirmed plurimodal projection to the taste cerebral cortex in the monkey.

The complex sensory image and problems of intermodality interactions (cf. Sect. 2.4, p. 160) appeared in Lawless et al. (Chem Senses 10(4):579–590, 1985).

Inhibitory responses (cf. Sect. 2.2, p. 144) have been also found by Contreras et al. (Physiol Behav 34:761–768, 1985) in the rabbit C.T.

Further interpretation of pattern coding (cf. p. 144), along with considerations about data treatment and choice of criteria, appeared in Grill and Erickson (Chem Senses 10(4):531–548, 1985), Erickson (Chem Senses 10(3):333–340, 1985), and Maes (Chem Senses 10(1):23–34 and 35–44, 1985).

Referring to descending modulations (cf. Sect. 2.5.1, p. 161) Hayama et al. found that neural structures above the pre- and mid-collicular level have tonic inhibitory or facilitatory effects on the response properties of NTS units (Exp Brain Res 60:235–242, 1985).

Lasiter et al. (Physiol Behav 35:277–285, 1985) confirmed that the anterior insular gustatory neocortex in rats is necessary for taste aversion learning, but not for taste reactivity or taste discrimination (cf. Sect. 2.5.2, p. 161).

Nejad (Chem Senses 11(3):283–294, 1986) showed better sensitivity to sugars on the rat palate than on the rat tongue. Responses to thaumatin (cf. p. 165) should be searched for in the greater superficial petrosal nerve.

Certainly, authors are presently increasing the number of stimuli applied (cf. Sect. 3.3, p. 171) when studying neural responses. Gill and Erickson (1985) used 32 stimuli; Iwasaki et al. 19 amino acids (Physiol Behav 34:531–542, 1985).

Gustatory reaction times measured by Kuznicki and Turner (Chem Senses 11(2):183–201, 1986) showed that different taste onset times helped human subjects in analyzing the components of a mixture. Yamamoto et al. (Physiol Behav 35:411–415, 1985) showed that reaction time measurements allowed experimental discrimination among various sweeteners (cf. p. 146). Interspecies differences (cf. Sect. 3.1.1, p. 167) were found also in the reaction to gymnemic acid inhibition by Hellekant et al. (Acta Physiol Scand 124:399–408, 1985).

Concerning monosodium glutamate and umami taste (cf. p. 168), refer to "Umami: a basic taste" (1985), Kawamura Y, Klare M (eds). Dekker, New York.

Referring to the number of receptor mechanisms (Sect. 3.2.4, p. 170) see Faurion (1985) in: Kawamura Y, Kate M (eds) Umami: a basic taste. Dekker, New York.

Subject Index

GPSR Compliance
The European Union's (EU) General Product Safety Regulation (GPSR) is a set of rules that requires consumer products to be safe and our obligations to ensure this.

If you have any concerns about our products, you can contact us on

ProductSafety@springernature.com

In case Publisher is established outside the EU, the EU authorized representative is:

Springer Nature Customer Service Center GmbH
Europaplatz 3
69115 Heidelberg, Germany

www.ingramcontent.com/pod-product-compliance
Lightning Source LLC
Chambersburg PA
CBHW061210240926
51846CB00006B/1172
* 9 7 8 3 6 4 2 7 1 0 6 2 9 *